NUCLEAR ENGINEERING FOR AN UNCERTAIN FUTURE

NUCLEAR ENGINEERING FOR AN UNCERTAIN FUTURE

International Symposium on the 20th Anniversary of the
Department of Nuclear Engineering, University of Tokyo

Edited by

Keichi OSHIMA
Yoshitsugu MISHIMA
Yoshio ANDO

PLENUM PRESS, NEW YORK

Published jointly by
UNIVERSITY OF TOKYO PRESS
and
PLENUM PRESS, NEW YORK

ISBN-13: 978-1-4684-4186-4   e-ISBN-13: 978-1-4684-4184-0
DOI: 10.1007/978-1-4684-4184-0

Library of Congress Catalog Card Number   81-84687

Plenum Press is an imprint of
Plenum Publishing Corporation
233 Spring Street
New York, N.Y. 10013

Contents

Organizing Committee
Preface
Opening Address
Opening Remarks

Part V. Research and Development of Fusion Technologies

Part VI. Nuclear Engineering and Technological Innovation

Part VII. Conclusion

Organizing Committee

Honorary Chairmen
>   Takashi MUKAIBO
>>   President, University of Tokyo
>>   Professor (1958–1964), Department of Nuclear Engineering

>   Osamu NISHINO
>>   Professor Emeritus,
>>   Department of Nuclear Engineering

>   Akira OYAMA
>>   Director, Power Reactor and Nuclear Fuel Corp.
>>   Professor (1958–1969), Department of Nuclear Engineering

>   Yutaka YAMAMOTO
>>   Professor Emeritus,
>>   Department of Nuclear Engineering

Chairman
>   Keichi OSHIMA
>>   Professor,
>>   Department of Nuclear Engineering

Vice-chairmen
>   Yoshio ANDO
>>   Professor,
>>   Department of Nuclear Engineering

>   Yoshitsugu MISHIMA
>>   Professor,
>>   Department of Nuclear Engineering

Members
>   Mamoru AKIYAMA
>>   Professor,
>>   Department of Nuclear Engineering

>   Shigehiro AN
>>   Professor,
>>   Nuclear Engineering Research Laboratory

Akira SEKIGUCHI
Professor,
Department of Nuclear Engineering

Atsuyuki SUZUKI
Associate Professor,
Department of Nuclear Engineering

Yoneho TABATA
Professor,
Nuclear Engineering Research Laboratory

Yoichi TAKAHASHI
Professor,
Nuclear Engineering Research Laboratory

Takaaki TAMURA
Professor,
Nuclear Engineering Research Laboratory

Yasumasa TOGO
Professor,
Department of Nuclear Engineering

Taijiro UCHIDA
Professor,
Department of Nuclear Engineering

Hiroaki WAKABAYASHI
Associate Professor,
Nuclear Engineering Research Laboratory

Genki YAGAWA
Associate Professor,
Department of Nuclear Engineering

Michio YAMAWAKI
Associate Professor,
Research Center for Nuclear Science and Technology

Secretariat
Hiroko BABA, Noriko KANZAKI, Masakazu OKAYASU,
Yoshiro YAMADA, Makiko YAMAMOTO

Preface

This is the official record of the International Symposium on "The Role of Nuclear Engineering for an Uncertain Future" which was held on November 5 and 6, 1980, at Keidanren Hall in Tokyo, in connection with the 20th Anniversary of the Nuclear Engineering Department, Faculty of Engineering, University of Tokyo.

Eight specialists from all over the world were invited to contribute papers to the symposium, and the professors of our Department presented a paper each. The Symposium was divided into seven sessions, chaired by professors of the Department according to their specialties.

About 200 scientists attended the symposium, and some of them joined the discussions. The symposium was fruitful and very successful from every point of view, and highly evaluated by the attendants as well as by concerned people outside.

This success is due to the successful organization and good performance of the staff of this symposium, to whom I would like to express my gratitude. I also hope that these proceedings will be useful to the specialists who are concerned with the uncertain future of nuclear engineering as well as with the role of Universities in that future.

March 1981

Yoshitsugu MISHIMA
Vice-chairman
The Organizing Committee

Opening Address

It is a great pleasure and honor for us to hold this International Symposium commemorating the 20th anniversary of the establishment of the Department of Nuclear Engineering of the University of Tokyo, with the participation of such distinguished guests involved in the development of nuclear energy from Japan and abroad. We are especially happy that many persons with whom we have had long and intimate relationships are here with us.

I would like to express my sincere gratitude to our guests from abroad who accepted our invitation in spite of their busy schedules and have come a long way to Japan. Also, I extend my heartfelt thanks to the individuals, industrial companies, and organizations which kindly contributed their financial support in response to our appeal for this symposium. This support has enabled us to organize a symposium even more splendid than we expected.

When we began planning the 20th anniversary of the Department, a proposal to hold an international symposium with the theme of "The Role of Nuclear Engineering for an Uncertain Future" came especially from young professors and staff members, and with the cooperation of the participants here it resulted in this symposium.

The 20th anniversary means that the Department was established in the 35th year of Showa, namely, 1960. Most of the professors and staff members who joined the Department have been involved more or less in the nuclear development of Japan since the 29th year of Showa, 1954, or even before. The Department has indeed been cooperating on the growth of nuclear energy in Japan till today. There have been many events, good and bad, but now the Japanese nuclear industry and technology have reached the level of high international competence with a sound foundation and Japan is playing a leading role internationally in this field. It is much more than what we could imagine in those early days. Also, today, the nuclear industry has been increasing its importance as an expected main source of energy supply for this country in the wake of the oil crisis.

On the other hand, however, we are confronting many problems such as international political relationships, the nuclear fuel cycle, and safety, and also the new technological developments of fast breeders, fusion, and so on. We have a strong feeling that nuclear energy is entering a new era. The past twenty years have seen the establishment of a sound basis for the peaceful use of nuclear energy. Now we have a new task: building on this base to make nuclear energy a reliable main stay for the world's energy production. We believe that whether nuclear energy can meet expectations and become the major supplier of energy in the future will greatly depend on the education and research in nuclear engineering at universities.

In this regard, we strongly look forward to the presentations and discussions of participants from home and abroad during the symposium to provide us with fruitful suggestions and perspectives for nuclear engineering education as we enter our third decade.

November 5, 1980

Keichi OSHIMA
Chairman of the Symposium

Opening Remarks

Mr. Chairman, Ladies and Gentlemen:

It is a great pleasure for me to address to you on the occasion of this symposium commemorating the 20th anniversary of the Department of Nuclear Engineering of my faculty. First I would like to welcome you and thank you for your participation, particularly those from overseas who have been our colleagues and friends of the Department for the past 20 years.

Now let's look back to the past briefly.

1955 was the year when budgeting for atomic energy development was started in this country. Soon afterward, several key organizations—the Japan Atomic Energy Commission, the Atomic Energy Bureau in the Science and Engineering Agency of the government, and the Japan Atomic Energy Research Institute—were established. The development and the growth of nuclear research and industry have been remarkable ever since. The number of nuclear power stations currently in operation number 22, and total power of 15 million KWe has been installed already.

The University of Tokyo also lost no time in recognizing the great need for the education and training of researchers and engineers in this field, devoting a great deal of discussion to the optimum system. In 1958 we set up several laboratories in our Faculty of Engineering, and the Department of Nuclear Engineering was formed in 1960. Over 500 students have graduated from this department since then. They have been active in nuclear industries, research institutes, and universities, playing important roles both in basic research and in developing nuclear technologies. This Department would have been unable to thrive as it did without your kind support and help.

Nuclear energy has so many ramifications in the economic and political spheres that delineating the direction of its future development is not a simple task. Since we need to overcome various problems in order to gain public acceptance of nuclear energy, I believe it is quite timely to hold this symposium at this time with the object of clarifying the role of nuclear engineering in an uncertain future. This will surely be a good opportunity, too, for us to convey our beliefs to you in that regard. I hope that this symposium can contribute to the sound development of nuclear technology in the world. I am sure we can find a path to the future soon. Thank you very much.

November 5, 1980

Yoshihiro HISAMATSU
Dean, Faculty of Engineering
University of Tokyo

Part I

International Cooperation on Energy

International Cooperation on Nuclear Energy

W. Kenneth DAVIS

1. Introduction

Some would think that now is not a time for celebration in the nuclear industry. However, I disagree. It is true that there are many problems confronting the industry, some of which are the subject of my talk. But I have always been an optimist, hopefully a realistic one; and further, as an engineer, I really believe that ultimately the facts will prevail even in a highly political environment.

The simple fact, which at least we all recognize, is that nuclear power is a vital, indeed absolutely essential source of energy which is not only vast in its potential but also safe, environmentally benign, and relatively inexpensive. While it cannot, at least in the short term, solve our problem of providing enough transport fuels and those of needs for small energy sources, it can provide essentially limitless amounts of electric power and large-scale sources of heat.

It is my conviction that we are really at a point where we all can take pride in our past accomplishments and look forward enthusiastically and confidently to the future despite having to overcome a few more problems on the road ahead—which, after all, is what makes life challenging and interesting.

Nuclear power is producing 13 percent of your electricity in Japan, 11 percent of our electricity in the United States (over 25 percent in 11 states of the U.S. and over 50 percent in 2 of those states), and similar fractions in other industrialized countries—and the fraction is growing.

With that as a prologue, I would like to discuss briefly the worldwide energy situation and prospects, the relevant history of the development of nuclear power in the international context, the present international and U.S. national nuclear situation, and my view of the necessity for turning what seems to have become an adversary situation back into one of fruit-

---

Vice President, Bechtel Power Corporation, U.S.A.

ful international cooperation, especially in the nuclear area.

2. World Energy Situation

An appreciation of the essential nature of the role of nuclear energy rests on an understanding of the worldwide energy situation. Table 1 shows the current pattern of worldwide energy use and the estimated ultimate recoverable energy resources of various types.[†]

Table 1. World energy consumption (1978) and estimated ultimately recoverable resources.

Energy source	Consumption		Estimated ultimately recoverable resources	
	quads/yr	%	quads	%
Conventional oil*	127.7	47.6	13,750	3.0
Natural gas	54.0	20.1	9,945	2.2
Coal	77.9	29.1	140,000	30.9
Nuclear[2*]	6.3	2.4	269,600[4*] (4010)[5*]	59.5
Oil shale	2.0	0.7	14,600	3.2
Tar sands and heavy oils	0.2	0.1	5,010	1.1
Total (rounded)	268	100	453,000	100
Renewables[3*]	17.1			
Total	285			

* Includes natural gas liquids; [2*] resources are for WOCA; [3*] hydro only; [4*] based on use in LMFBRs, considering U_3O_8 at "forward costs" up to $50/lb; [5*] number in parentheses assumes U and Pu recycle in thermal reactors.
[†] From "Survey of Energy Resources—1980" (World Energy Conference).

While it is evident that there are still substantial amounts of oil and gas yet to be discovered and produced even in terms of present consumption rates, it is obvious that there is an enormous disparity between the present pattern of use and the resources available in the long run. Further, the location of the oil and gas resources is generally disadvantageous in terms of political control, transport costs, and prices to many of the industrialized consuming nations, especially Japan.

The present pattern of use is dictated to a large extent by the necessities of transport use, but also by previous generally low prices and high availability. It will be difficult to change the use pattern and, in any event, major changes will require decades, not years. However, there is, in the long run, no alternative to such change.

The United States has a pattern of use versus availability of resources which is even more extreme, as shown in Table 2. The extreme dependence on oil (48% of total use) is supported by importation of over 40% of the amount used. It appears unlikely that such imports will decrease significantly during the next few years, and even then only if vigorous measures are successfully taken to increase domestic production, develop oil shale and oil from coal, and make more sparing and efficient use of the oil available (already stimulated by the large increases in oil prices).

Table 2. U.S. energy consumption (1979) and estimated ultimately recoverable resources.

	Consumption		Estimated ultimately recoverable resources	
Energy source	quads/yr	%	quads	%
Conventional oil*	36.3	48.1	921	0.6
Natural gas	20.5	27.2	1,500	1.0
Coal	15.1	20.0	38,000	24.2
Nuclear	2.75	3.6	105,000[2]* (1562)[3]*	66.9
Oil shale	0	0	11,370	7.2
Tar sands and heavy oils	0.8	1.1	102	0.1
Total (rounded)	75.4	100	157,000	100
Renewables	3.25			
Total	78.7			

* Includes natural gas liquids.
[2]* Based on use in LMFBRs, considering U_3O_8 at "forward costs" up to $50/lb.
[3]* Number in parentheses assumes U and Pu recycle in thermal reactors.

Even though worldwide recoverable conventional oil resources, as compared with current use, would give an R/P ratio (Resource/Production ratio) of 100 to 110 years, this cannot be considered at all reassuring. In past years the prevailing estimates were for a peaking of worldwide oil production (similar to the 1970 peak in the United States) around the year 2000 at a level of about 100 MBPD. These have been reduced in both time span and amount to a potential peak between 1985 and 1990 and overall production levels of perhaps 75 to 80 MBPD. Some believe that the present production level of about 63 MBPD will never be increased. While it is likely that production could be substantially increased from the technical point of view (assuming anticipated reserves could be discovered and developed rapidly), political factors along with the high investment cost of increased production rates will probably lead to the lower range of production level estimates.

The overall energy consumption of the developing countries will need to rise rapidly, as shown in Figure 1, if they are to continue to maintain a reasonable rate of economic development. Much of that increase is assumed

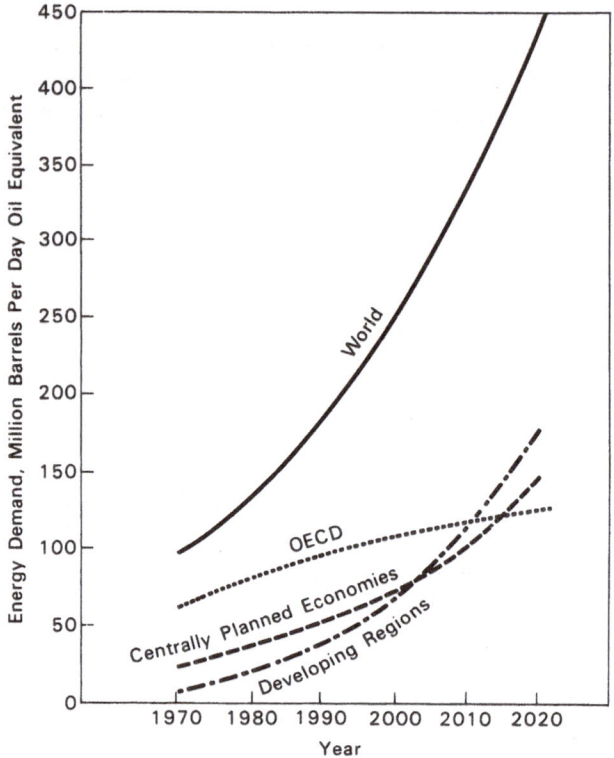

Fig. 1. The Conservation Commission's alternative scenario for demand.

to come from petroleum. The situation is clearly becoming critical even if we discount the enormous impact of possible oil cutoffs due to potential political/military events in the Middle East or elsewhere.

The industrialized countries cover a wide range of energy resource availability, from Japan which has almost none to countries such as the U.S. and the U.S.S.R., which have large indigenous resources. It is quite understandable that many of the developing countries (and some of the developed ones as well) are resentful that countries such as the U.S. are failing to develop adequately their domestic reserves of coal and uranium and are continuing to compete on the world market for what are almost sure to be decreasing supplies of export oil—and, what is worse, driving the prices even higher. When this is combined with the parallel concerns by the developing countries over availability of technology, I believe we are confronted with an increasingly serious and very broad international controversy.

Table 3. Potential world primary energy production in million barrels per day oil equivalent.

Resources	1972	1985	2000	2020
Coal	29.6	51.5	76.2	116.1
Oil	51.5	96.8	87.4	47.5
Gas	20.6	34.5	64.1	56.0
Nuclear	0.9	10.3	19.7	69.4
Hydraulic	6.3	10.8	15.2	25.1
Renewable, solar, geothermal, and biomass	11.6	14.8	25.1	44.8
Unconventional oil and gas	0	~0	1.8	17.9
Total	120.5	218.7	289.5	376.8
Demand	120.5	162.	255.	448.

Source: World Energy Conference; 5.8×10^4 Btu/bbl. oil equivalent, August 1980.

Table 3, which is based on the 1977 World Energy Conference study, shows a projection of worldwide energy production and anticipated requirements. The estimates of production are essentially the highest plausible rates except for nuclear power which is shown here at one-half the WEC projections—but now believed to be more likely (corresponding to 370 GW in 1985, 710 GW in 2000, and 2500 GW in 2020).

At the recent WEC in Munich, the maximum oil production rate was forecast at about 79 MBPD in 1990, roughly 20 MBPD less than shown. It is considered that, due to potential delays and shortfalls as well as geographical maldistribution, a margin between estimated future supply and demand of less than 15 to 20 percent could be critical.

The growing importance of nuclear power is clear, particularly for the year 2000 and beyond. While nuclear power is not the total solution (and I could spend a lot of time discussing coal, but will not), it is a most important part of it, and we should be considered irresponsible, and would be, if we failed to make use of nuclear power to the extent possible.

3. Nuclear Industry Development

Around 1950 active programs to develop nuclear electric power emerged in the U.S., the U.K., and the U.S.S.R.—followed a few years later by programs in Sweden, France, West Germany, Japan, and other countries. While a really vast number of potential power reactor ideas were suggested and evaluated and a significant number experimented with, the basic types utilized in each country ended up being adaptations of reactor technology developed for research, military, or production purposes—the gas-cooled reactor in the U.K. and France, the light water reactor in the U.S. and the U.S.S.R., the U.S.S.R.'s gas-graphite reactor, and the heavy water reactor in Canada and Sweden. The liquid meta-cooled breeder was a basic type considered in the early days in the U.S., the U.K., France, and the U.S.S.R., but in all cases the development, at least on a large scale, was deferred in favor of the simpler thermal reactor concepts which, it was thought, could be developed more easily and quicker, and had better (although very difficult) prospects for becoming economical.

The problems of proliferation were not discovered by the present administration in the U.S., but along with those of reactor safety were paramount considerations from the very beginning. The safeguarding of fissionable materials and accountability for them, as well as the development of bilateral, multilateral, and international arrangements and institutions, were primary elements in the early U.S. reactor development policies as well as those of other countries.

It was usually considered that, while reactors themselves, of almost any type, along with most fuel fabrication activities, were fairly easily controlled and inspected, uranium enrichment and fuel reprocessing plants would be located in large centralized facilities with very careful control, inspection, and monitoring—and would be few enough in number so that this could be achieved at a reasonable cost.

It was a basic tenet, still shared by most of us, that fuel reprocessing and recycling was desirable to achieve the best economics, essential to make prudent and responsible use of our uranium resources, and, of course, fundamental for developing the breeder reactor, the foundation for the long-term future of nuclear energy.

It was also a basic concept of the pioneers in nuclear power that it is a technology of value to the whole world—that its benefits should be shared with the world and that its development would be speeded by international cooperation and collaboration. It was also believed that such international cooperation would lead to effective control of the proliferation potential since, under such circumstances, this would be in everyone's best interests. It was also recognized that there were (and are) no ways of achieving absolute control—only ways of minimizing and deferring the risks. Can we today say that anything has changed to invalidate these concepts? I do not think so.

As a step in this direction, the first International Conference on the Peaceful Uses of Atomic Energy was held in Geneva in 1955. This was accompanied by large-scale declassification of power reactor technologies in the U.S. and other countries. In the U.S. this was facilitated by passage of the Atomic Energy Act of 1954 which permitted private ownership of power reactors.

This was followed by other major initiatives in which the U.S. took a major part, along with other nations of the same mind. There was a series of

Geneva Conferences in 1958, 1964, 1971, and 1974 (in Salzburg) following the 1955 Conference. In December 1953, President Eisenhower presented the Atoms for Peace Program at the United Nations. The International Atomic Energy Agency (IAEA) was formed in 1957 with broad international support and increasing responsibilities. The nuclear Non-Proliferation Treaty (NPT) emerged from the IAEA and has gradually gained support by promising nuclear technology and assistance for safeguards against proliferation. The U.S. and other countries entered into a variety of bilateral and multilateral agreements, not only for the exchange of research and development results but also for the supply of nuclear fuel, equipment, and services—all with a view to expanding the beneficial uses of atomic energy on a mutually satisfactory basis and, as much as possible, through normal industrial channels.

In this regard, Japan was the first nation (1962–63) to negotiate with the United States for the transfer of bilateral safeguards to the IAEA. This involved acceptance of IAEA inspectors. It reflected Japan's commitment to the peaceful use of nuclear energy as stated in the Japanese Constitution.

With regard to costs, nuclear power has achieved economics which are very competitive with other ways of producing electric power. Table 4 shows the results of a comparison we recently made for a typical mid-U.S. location, which is generally valid in other areas of the U.S. as well

Table 4.   Levelized generating costs (mills/kwhr).

Plant size	2 × 1200 MWe		3 × 800 MWe		
Fuel	Nuclear (LWR)	High-S coal with scrubbers	Low-S coal without scrubbers	Low-S coal with scrubbers	Low-S oil without scrubbers
Levelized costs (unescalated)					
Fixed charges @ 9.4%	14.1*	10.8	9.6	11.8	7.5
Fuel	6.4²*	15.6	16.2	17.0	56.5
Operating and maintenance	2.3	6.3	2.7	3.8	1.9
Total (1980 $)	22.8	32.7	28.5	32.6	65.9
Levelized costs (escalated)					
Fixed charges @ 17%	46.3	36.4	32.4	39.7	24.5
Fuel	16.7	38.1	39.5	41.5	133.0
Operating and maintenance	5.6	15.1	6.4	9.0	4.6
Total (1990 $)	68.6	89.6	78.3	90.2	162.1

* Includes sinking fund charges to cover plant decommissioning.
²* Assumes reprocessing and recycle of U and Pu. Add about 10% for once-through fuel use without reprocessing.

as much of the rest of the world. I do not believe it is necessary to convince this audience that nuclear power is safe (especially in view of the Three Mile Island Unit 2 incident) or environmentally benign. In any case, there are others here today who can argue these points more knowledgeably and persuasively than I can.

4. Necessity for International Cooperation

Energy demand, supply, and development are matters of increasing international concern and are becoming critical, if not dominant, in international relations. This is a relatively new area, especially to the world's diplomats and politicians, and needs to be approached carefully, patiently, and with great understanding. New ideas will surely be needed but will not be identified or accepted overnight. We have countries falling into different categories: the petroleum-rich versus the petroleum-poor, the energy resource-rich versus the energy resource-poor, and, most importantly, the developing countries versus the developed countries. The situation is complex and growing more so. We must arrive at workable solutions to provide adequate supplies of energy in usable form for all countries to sustain desirable living standards and enable those countries which do not have them to achieve them. The penalty for failure may be extreme.

Nowhere is this problem more evident than in the nuclear power area. Atomic energy, which poses the greatest threat to mankind, also holds out the greatest promise. However, real and continued international collaboration and trust are essential to hold back the threat and achieve the promise. This was the vision of 35 years ago—and is as true today as it was then.

Many of the normal commercial aspects of the development of nuclear power have been pursued successfully, including the supply of fuel, services, equipment, and licensing arrangements which are important between industries in many countries, including Japan and the United States. These have been carried out under appropriate bilateral, multilateral, or international intergovernmental agreements.

The number of companies seeking to supply nuclear steam supply systems worldwide probably exceeds the number which can operate profitably. The formation of multinational groups and consortia of one sort or another appears not only inevitable but desirable. In the area of uranium enrichment and fuel reprocessing, this would also appear to be the logical outcome. These developments can also be expected to be intertwined with the development of international safeguards institutions.

The breeder reactor appears to be an interesting case in itself. The cost of development to commercial operation will be very high (including the early models which will not be competitive). The cost of establishing the fuel cycle on an economic basis may, in fact, be even greater and is, as yet, a problem which has not really been thought through, much less resolved. In the end, the breeder reactors are likely to be very large per unit in order to be economic (say 1,800 to 2,000 MW) and built in "clusters" to allow on-site processing and fuel refabrication to cut down plutonium inventory and ease the safeguards problem. Thus they will only be of interest to countries with a large base load and a well-integrated grid. Other countries may use thermal reactors and produce fuel for the breeders. Thus the market for breeders may be not only limited but circumscribed by safeguards requirements.

Some of us have speculated on the number of viable NSSS (Nuclear Steam Supply System) suppliers in such circumstances, and the result is usually 3 or 4. I believe that what will emerge is perhaps 3 or 4 groups or consortia supplying fast breeder reactor steam supply systems. These will likely be multinational groups based on the international groups formed to carry out the large-scale developments—which can probably only be financed on that basis anyway.

To a large extent, the problems of Japan and the United States are the same although the causes are different—how to secure a major place in the development groups and then in the supply consortia. Japan is trying to catch up and establish a position. The United States, which was clearly ahead, has for various reasons abandoned the leadership and may find it difficult to regain in time to secure an advantageous position for its industry.

The problems involved in the engineering development and commercial deployment of fusion are almost impossible to visualize at this time. However, it seems clear that the cost will be enormous and the time involved will be very long. It is difficult to imagine how this could be done successfully without a continuation of the past international scientific cooperation in fusion in engineering and commercial development as well. I doubt that it would be feasible without it. Steps must be taken to accomplish this transition from the scientific effort in fusion to the engineering effort which should start almost at once.

It should be noted that the investment in new power generation facilities represents the largest single class of industrial investment in the United States and most other industrialized countries, including Japan, and requires major interaction with worldwide financial institutions. Facilities for synthetic fuels may equal or exceed such investments in time in some countries, but the financial needs for electric power will continue to be very large for the foreseeable future.

It seems evident that the future of nuclear power lies in the direction of further international cooperation both at the industrial level and at the national and international level.

5. Current Nuclear Situation

Despite the clearly advantageous factual case for nuclear power, the program is in critical shape in the United States and is having a variety of difficulties in many other countries which need to use nuclear power. The effect of the difficulties in the U.S. can be illustrated by Figure 2, which shows power reactor capacity built, under construction, and ordered since 1965. The total has declined since the end of 1974, which coincided with the abolition of the U.S. Atomic Energy Commission and the U.S. Joint Congressional Committee on Atomic Energy. Other significant events were the cancellation of the government program and regulations on high-level waste disposal in early 1975 and the "deferral" of fuel reprocessing by President Ford in 1976. This was followed by an active policy of non-sup-

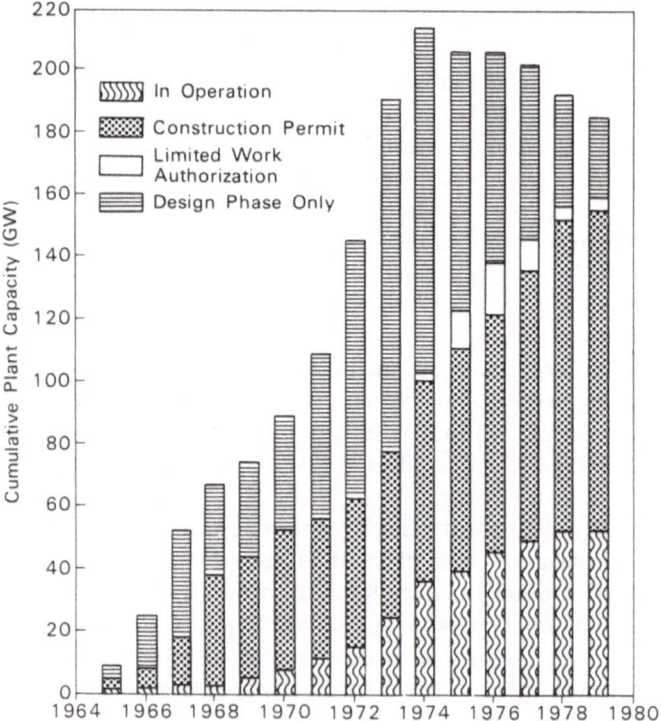

Fig. 2.   U.S. nuclear power plant status, 1965 to 1979.

port of nuclear power by President Carter in 1977, including abolition of fuel reprocessing and measures to reduce or eliminate work on breeders. The effect has been to permit interminable intervention in licensing proceedings by those who are more interested in destroying our sociopolitical system than in the safety of nuclear power, and a vast proliferation of licensing requirements and changes.

In addition, the sudden surge in oil prices and rapid escalation in the prices of other fuels as well as plant investment costs have driven electric utility costs up very rapidly since 1974. The rate and revenue regulation by our 50 state utility regulatory agencies has generally failed completely to keep up with these increases. As a result, very few of our utilities are able to order a new plant, as badly as it might be needed, and in addition have had to slow down, defer, or cancel many projects already under way.

The unilateral non-proliferation policy initiated by President Carter in April 1977 and compounded by the Nuclear Non-Proliferation Act of 1978 has proven disastrous for the world nuclear industry, for U.S. export business, and for international business relationships, and has severely damaged U.S. worldwide credibility and leadership in the nuclear field. It is generally considered that, in addition, this policy has *increased* significantly the risks of proliferation. Despite the clear results of the International Nuclear Fuel Cycle Evaluation (INFCE) which was suggested by the U.S. in response to criticism of its policy, there is little evidence of any change in policy or practice by the present U.S. Administration. The INFCE results demonstrated that although some technical improvements were feasible, the proliferation risks of the conventional reactors, breeders, and their fuel cycles could be dealt with effectively, but only by strengthened international institutions and controls—an approach which demands international cooperation and leadership.

The impact of these events on many of our partners in the nuclear field has been severe and has led to a sad state of strained relations at a time when we should be working together to solve all of the problems of nuclear power including the risks of proliferation. This is probably evident in Japan more than almost anywhere else in the world. The problems simply cannot be resolved by the adversary intergovernmental relationships which have evolved. Real cooperation based on mutual understanding, credibility, and trust must be re-established between the U.S. and the governments involved, as well as among members of the academic and industrial sectors.

These are, perhaps unfortunately, not the kinds of problems which can be solved by the academic and industrial communities. Rather, the solutions must start at the top with our nations' leaders. Positive direction must start at the top and be continuous to be effective.

6. Prospects for the Future

What about the future of nuclear power? Why am I still an optimist about nuclear power making a growing contribution to our national and international energy supplies? Why do I believe the problems can be solved— and if so, how?

Let me try to address these questions even though this is an inauspicious moment to speculate, with the results of the elections in the United States available only a few hours from now.

One way to approach the questions is to hypothesize a series of actions which, I believe, if taken, would lead to the renewed viability of nuclear power. Here are some of the types of steps which might be taken in the United States:

The Administration states positive, unequivocal support for nuclear power.

This policy is accepted by the key federal and state agencies.

A separate nuclear agency is established to deal with the production and power-related nuclear programs (excluding licensing and regulation), including thermal reactors, breeder reactors, fusion, and fuel cycle facilities. New Congressional oversight might be established.

The Nuclear Regulatory Commission is restructured and redirected to provide prompt and orderly nuclear licensing.

The Administration (and the Congress) revise the proliferation policy taking into account the findings of the INFCE Study and take the lead in seeking to expand the role of the IAEA, the scope of the NPT, and the strengthening of international safeguarding programs and institutions.

The Administration drops its opposition to nuclear fuel reprocessing and the breeder. It supports continuation of Clinch River and takes steps to participate in international collaboration in the construction of commercial prototype breeder reactors.

The Administration acknowledges responsibility for nuclear waste storage and announces a positive program for early demonstration of high-level waste storage.

The Administration takes leadership in assuring the clean-up and restoration of Three Mile Island Unit 2.

Federal and state actions are taken to provide electric utilities with an adequate basis for financing new facilities.

Your first reaction might be about public opinion in the United States. I would suggest that public opinion is still generally in favor of proceeding with nuclear power, that public opinion would be swayed to a large extent by a positive attitude on the part of the government, and that later, if not sooner, the public will come to realize that there are only limited solutions, including nuclear power, to sustaining or improving their standard of living.

My concern is over the timing of the kinds of actions outlined above since I feel they must come in time. I hope we will have the leadership to bring them about in an orderly and timely way. However, I am convinced that ultimately they will come.

Such a set of initiatives and changes in the United States would surely lead to changes in other parts of the world as well. It not only would set the stage for new political and institutional cooperation among governments but also would facilitate the growth of international cooperation in the business world. As noted earlier, this is essential for a healthy worldwide nuclear industry. The prospects for regional and multinational enrichment and processing facilities along with waste management would be opened up for realistic consideration and action. The development and deployment of the breeder reactor when needed would be assured. The road to commercial fusion, when it is timely, would be much broader and more likely to see early success.

This is a dream, if you will, of a new nuclear era. It is based only on logic and facts. The facts are the economics, safety, and vast potential of nuclear power under adequate international controls.

I believe that logic and facts will prevail—and hope that you share that conviction with me.

Nuclear Energy and International Cooperation

Keichi Oshima

1. Introduction

There is no need to emphasize to this audience that nuclear energy cannot be developed without international cooperation at either the industrial or the academic level. However, as Dr. W. K. Davis pointed out, there have been some marked political, economic, and social changes in recent years which are posing constraints to international cooperation in nuclear energy.

One of the most recent issues of concern has been the non-proliferation policy of the United States during the Carter years. As all of us who have been involved in nuclear energy know well, the non-proliferation of nuclear weapons and peaceful use of nuclear energy have been the two basic concerns of the nuclear engineering community from the start of nuclear power development. International cooperation in nuclear energy has always had to take into consideration the problem of non-proliferation. Through bilateral agreements, IAEA (International Atomic Energy Agency) safeguards, and finally with the Non-Proliferation Treaty, these two requirements were thought to be successfully balanced. Everyone has known that there is no perfect solution for these matters and that the balance achieved is a delicate and fragile one. However, in any case, non-proliferation is a political matter which cannot be solved completely by technological solutions.

The second constraint to nuclear power development of an international character is the environmental issue. Opposition by environmental groups is now organized internationally and there is a strong movement against waste disposal and waste management on a global scale. For example, shipment of Japanese nuclear spent fuel to Europe was opposed by a group in the U.S., and the cargo boat was not permitted to stop there.

Also, since the Three Mile Island nuclear accident in the U.S., stricter safety requirements and regulations on siting have been demanded by the public in other countries as well, and again international interaction has been very strong.

Department of Nuclear Engineering, University of Tokyo, Tokyo, Japan.

The third constraints is not very evident yet, but after the oil crisis, which caused friction and imbalances in international trade, sentiment in restrictions on the transfer of advanced technology to competitor countries has arisen. This has been complemented by political consideration concerning the non-proliferation of sensitive nuclear technology, and as a consequence there seem to be growing constraints on international cooperation in the field of nuclear technology, especially at the industrial level.

As a nuclear engineer I believe that nuclear technology is now well established and also that, under the new economic situation prevailing since the onset of the oil crisis characterized by rapid oil price increases and supply uncertainties, nuclear power is the most advantageous alternative to oil as a source of energy in terms of its cost and security of supply.

I recall a statement made on one occasion by Dr. Irving Weinberg that it is fortunate for mankind to have nuclear power ready in our hand now at this time of oil crisis; all the obstacles to the use of nuclear power are politicial, social, or institutional, and they can be solved immediately if once the public decides to do so. It would have taken several decades to develop nuclear technology if we were only starting now.

However, the situation of nuclear power programs in most countries is not so optimistic as expected. Even though at summit meetings political leaders make statements expressing their agreement to establish oil import ceilings and to promote nuclear power, in reality most countries, except France, are suffering delays in their programs.

I believe it is a challenge for nuclear engineers as a worldwide group to overcome the present situation and put nuclear programs back on the track. I learned that even for a country like the U. S., endowed with abundant natural resources of fossil fuel, nuclear power is a most important energy source for the future. For Japan, without domestic energy resources and almost completely dependent on oil imports, the development of a nuclear power program is vital for its industry and well-being.

The problems and constraints impeding our nuclear power program cannot be overcome by only one nation; international cooperation with common efforts to solve the problems is essential.

2. History of Nuclear Energy in Japan

The development of nuclear energy in Japan was started with international cooperation. In 1953, President Eisenhower made the famous Atoms for Peace declaration at the General Assembly of the United Nations, and its impact reached Japan in the next year; there were heated discussions in academic circles, especially at the National Research Council, of whether Japan should join the U. S. proposal for the peaceful use of nuclear energy, which can be used for military purposes on some occasions. Political movement in the Diet was much quicker, and Japan's first nuclear energy program was launched with a budget of ¥25 million during the 1954 Diet session. A study mission was sent overseas, and I believe Prof. A. Oyama was the first Japanese student of nuclear engineering to attend the Argonne Reactor School.

If I may, I would like to refer to my own experience in the early 1950s. As a young researcher in applied physical chemistry, nuclear energy attracted my professional interest so strongly that I tried to assemble all the information available on nuclear reactors, even though research on nuclear energy was not allowed in Japan during the U.S. occupation. I think the first information on nuclear reactor design that I saw was of the French research reactor at Saclay. Our group at the Institute for Science and Technology, University of Tokyo, which was a reconstruction of the Institute of Aeronautical Science banned by the U.S. authorities, started a study on the production of heavy water with the collaboration of Mr. K. Nakamura, head of the Kawasaki Factory of Showa Denko. Heavy water was produced by making a cascade of electrolytic cells for hydrogen production. This was before the official approval of the Japanese nuclear energy program. When the program started in 1954 I remember that Prof. N. Kameyama, whose student I was, approached Showa Denko asking for cooperation with his new nuclear energy research group at the Institute of Physical and Chemical Research (where he was the president at that time), and to his surprise he found out that they were already working secretly with us. Anyhow, with the start of nuclear energy research, information flooded into Japan and we all sought international cooperation.

In 1956, the Japan Atomic Industrial Forum was established, and the U.S.-Japan Joint Atomic Industrial Forum Conference in 1957 was a big event for Japan. Many outstanding people involved in nuclear energy from the time of the Manhattan Project visited Japan and I was excited to meet and have personal contact with those people whom I only knew through publication; for example, I met Mr. P. J. Selak, the author of the review article on heavy water in *Chemical Engineering Progress*. I recall our enjoyable trip to Kyoto and Nara on that occasion with the participants of the Conference including Ashton J. O. Donnell who is now a colleague of Mr. W.K. Davis's at Bechtel. Really we felt that we became a part of the world nuclear community.

In 1958, many of us attended the 2nd Geneva Conference on Peaceful Uses of Nuclear Energy. Japan was a newcomer in nuclear energy, and Dr. H. J. Bhabha of India was a vice-president of the Conference, representing Asia's advanced country in nuclear energy. This conference was the first one to which Japan sent a large official delegation and presented papers on Japanese activities in nuclear energy. I delivered the speech of Commissioner I. Ishikawa of the Japanese Atomic Energy Commission on his behalf at the general session, and that was my first experience in presenting a paper at an official international conference of the UN. What impressed me most vividly was that now Japan was participating in international cooperation in nuclear energy.

Later that year I visited several other countries including the U.S. where I stayed at MIT and had the privilege of studying with Prof. M. Benedict, who is here today as a guest. At that time I happened to hear from Prof. W. G. Whitman, then the chairman of MIT's Chemical Engineering Department, the story of his organizing the First Geneva Conference in 1954. He first flew to Moscow to reach an agreement with the Soviet Union in collaborating with the U.S. on the presentation of papers on the peaceful uses of nuclear energy and with this understanding in mind he started the Conference. Thus, the peaceful use of nuclear energy started with international cooperation even between the West and the East. In 1957 this led to the creation of the IAEA as a UN organization for international cooperation toward the peaceful use of nuclear energy and toward security for the non-proliferation of nuclear weapons. Japan has fully cooperated with IAEA from its start until today.

It is interesting to note today that in those early days of peaceful uses of nuclear energy in the 1950s, already Japan was regarded by several authors to be the most suitable country for nuclear power. Energy prices paid by Japanese industry at the end of 1950 were about 1.00 to 1.30 yen per 1,000 kilocalories, based on domestic coal prices, compared to about 0.30 yen in the United States and about 0.80 to 0.90 yen in Europe, except Italy. Therefore, Japan was an industrial country with large energy demand, the highest energy costs in the world, and having an electricity network large enough to introduce large capacity power stations where nuclear energy can be competitive. In the 1960s, after the liberalization of imports of crude oil from the Middle East, Japanese energy prices fell rapidly to about 0.70 to 0.50 yen per 1,000 kilocalories, and one result was that the advantage for nuclear power has disappeared. Twenty years later, with the rapid increase in crude oil prices by a factor of twenty within a span of in seven years, Japan's energy situation has come back to what it was in the 1950s, and it is true that Japan is one of the countries in the world where nuclear power has tremendous advantages for the economy and in daily life.

The 1960s was a period of high economic growth and remarkable advancement of Japanese industrial technology. During this period, most of the staff members of our Department of Nuclear Engineering studied abroad; we sent many students, not only to the U. S., but also to the U. K., France, and other European countries. We have enjoyed close relations with the Department of Nuclear Engineering of MIT from the time that Prof. M. Benedict served as its first chairman, as all of you know. Our fast neutron source reactor "Yayoi" in the Tokai Nuclear Research Laboratory was built with help from the French CEA, following the design of Harmonie in Cadarashe

Nuclear engineering is an outstanding example of Japan's starting from nothing after World War II and catching up with American and European technology by learning from abroad and by our own domestic R & D efforts. As I mentioned already, international exchange and cooperation are nothing new in the field of nuclear energy and are taken for granted by everybody; we have counted ourselves a member of the international nuclear family. This feeling was not limited to the universities; national laboratories, industry, and even government agencies were all internationally minded.

Figure 1 shows how Japanese technology caught up with that of the Western countries in the 1960s. Japan, lowest in its level of technological

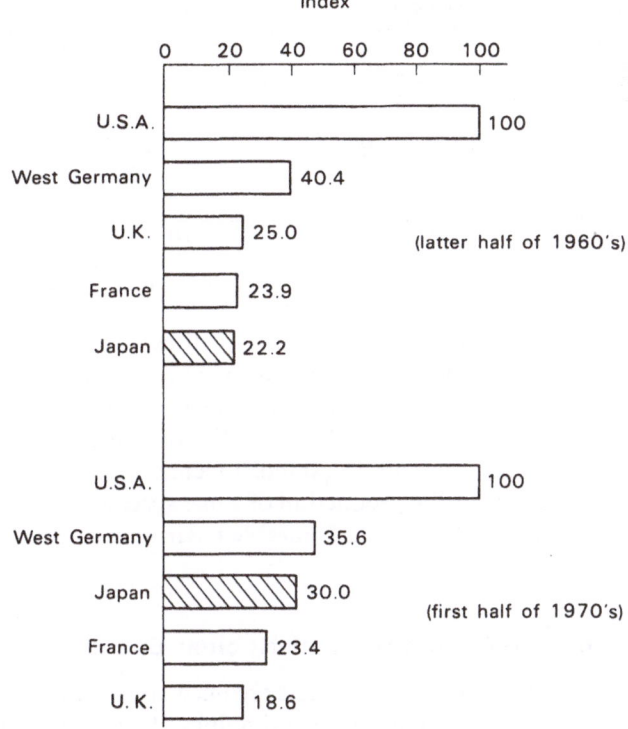

Fig. 1. Technological standards.

standards among the four major industrialized countries at the beginning of the 1960s, has narrowed the gap with the U.S. and moved into third place. Figure 2 compares the productivity of Japanese and U.S. industry. From 1955 to 1970, Japanese productivity grew by a factor of four while the United States dropped by about one-half. Being a "comprehensive technology" relying on many other industries for its components, there is no doubt that nuclear technology benefited from such overall advancements.

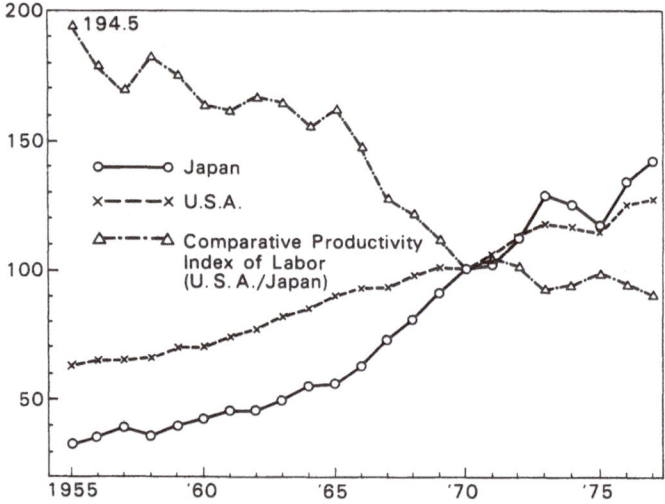

Fig. 2.   Productivity index of labor in manufacturing industries (1970 = 100).

Figures 3 and 4 show how Japanese industries combined imports of foreign technology from abroad with their own indigenous R & D efforts. They show that industries which paid more for imported technology have also spent more for their own R & D. It implies that import of technology was part of a general attempt to establish technological capability as a whole and not merely for the production of some specific products for sale. This was all possible only under a favorable environment of international cooperation.

3. Recent International Status of Nuclear Energy

Today, the international situation has changed considerably from those days of international cooperation in technology. In the 1970s the world economy entered into a period of slow economic growth, inflation, and

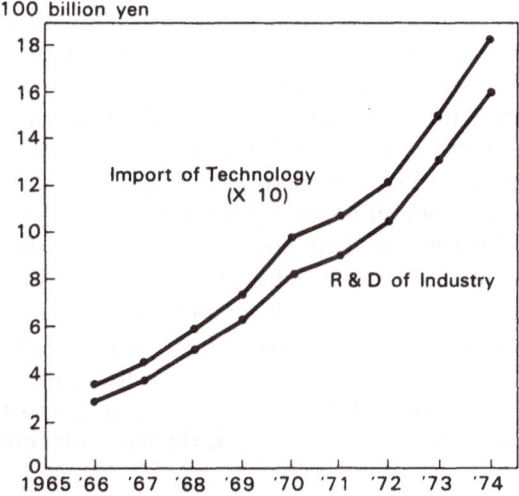

Fig. 3. Expenditures for import of technology and R & D by industry.

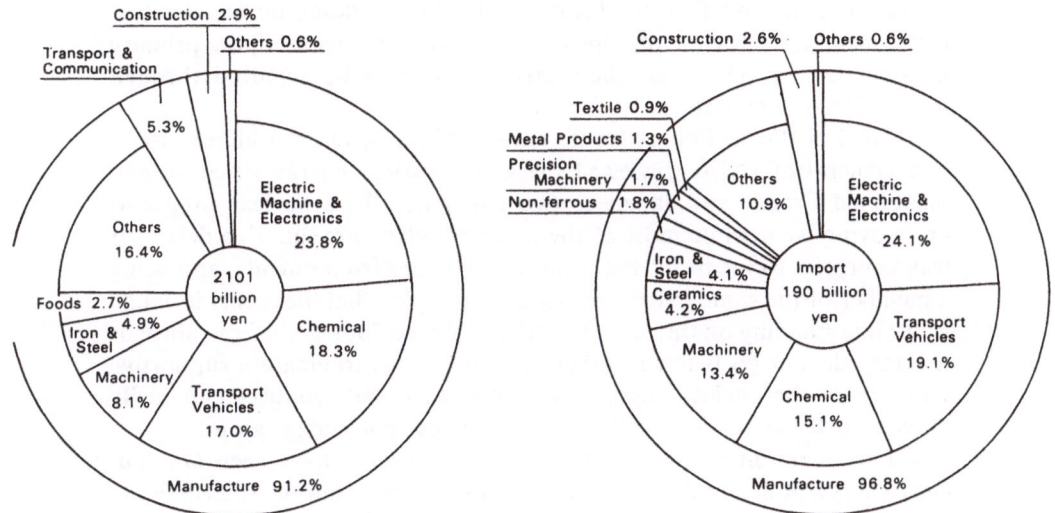

Fig. 4. Japanese industrial R & D expenditure and import of technology by sector (1977).

sociopolitical constraints. Since the oil crisis in 1973, the principles of free economy and free trade have been under pressure because of sociopolitical concerns about unemployment, decline of local industry, and balance of payments. Japan was able to manage the energy situation of rapidly rising crude oil prices relatively well because of its high productivity and techno- logical flexibility. However, this raised difficulties in some other OECD countries in the form of trade friction and also of resentment of Japan's easy acceptance of OPEC's crude oil price increases. Even though Japan may be able to pay high oil prices, other countries cannot afford it, and if Japan increases exports to these countries to earn foreign exchange for buying oil it means a double menace to these other OECD countries. Also, it is not desirable, from the security point of view, for Japan to rely so heavily for its energy supplies on Middle East oil in the light of the present uncertain political situation in that area. Therefore, for the sake of the world and in its own interest, Japan has to seek an alternative to oil, and there is no question that nuclear energy is the most advantageous alterna- tive.

In fact, Japan's nuclear power program is very substantial even after readjustment for the anticipated delays in construction. In 1990, it is ex- pected to reach 53 GW (gigawatts) installed capacity accounting for 10% of primary energy needs and 25% of total electricity generation capacity, compared with the present capacity of 12 GW which represented 12% of total electricity generation capacity in 1980. Also the development of a fast breeder reactor and the establishment of a closed domestic nuclear fuel cycle (including enrichment, reprocessing, and waste disposal) are primary national targets. However, these targets can only be achieved through international cooperation.

At the 8th World Energy Conference in Munich, Dr. S. Eklund, Secre- tary-General of IAEA, reviewed the world nuclear energy situation and mentioned France and Japan as two countries where nuclear programs are moving well, while most of the other countries are suffering delays. It was a surprise to me to see that things look rosier from outside, and being a member of the same panel I responded by saying that the reality in Japan is just like standing on thin ice; even though figures look good from abroad; if things do not go better in other countries so as to create a supportive atmosphere for nuclear energy, difficulties in other countries can easily move into Japan to disturb the whole Japanese nuclear program.

The most recent political issues for nuclear energy have been U.S. nu- clear energy policy under the Carter Administration and the International Fuel Cycle Evaluation (INFCE). As an irony of history, the non-prolif- eration argument was raised when nuclear power had just begun to play the most important role as an alternative energy source. Political discussions have been continuing for two years, and though it seemed to be a tedious task when nuclear power development was so urgent, I believe that INFCE contributed to future nuclear development not only because non-proliferation is an essential issue in the peaceful use of nuclear energy, which has to be settled by worldwide political common agreement, but also because it provided a good opportunity for participating countries to increase their understanding of each other's economic and political situations with regard to nuclear energy. Some people in Japan compared it with the textile negotiations but it seems to me quite different because at INFCE there were more constructive and well-coordinated arguments. One (and perhaps the most important) reason for this was that in the nuclear energy community there existed a long history of international cooperation and thus most of the participants in the meetings, at least the technical people, have known each other for many years. In nuclear energy, there was an international cooperative group from the start at the Manhattan Project, and when it moved to peaceful use at the Geneva Conferences and other forums, the group was joined by countries like Japan and West Germany to be expanded to a real international community. The existence of such a community forms the basis of an international joint effort for development of peaceful uses of nuclear energy for mankind.

4. Conclusions

As described above, international cooperation is an inherent characteristic of nuclear development. Nuclear energy started with international cooperation and has been developed by it. Also, nuclear energy is different from fossil energy resources in that it is highly technology-intensive while others are resource-intensive. Until nuclear technology was developed uranium was never regarded as an energy resource: it is different from coal or oil which can be burnt directly. Today, for light water reactors, the cost of natural uranium is less than 10% of the total power generation cost, while for oil-burning power plants the cost of fuel averages about 70%. This means that nuclear electricity cost is 7 times less sensitive to changes in the cost of fuel than is oil-fired electricity. With the progress in nuclear technology, the thermal efficiency of natural uranium can be increased about a hundred times more by the introduction of fast breeder reactors and the cost of resources is almost negligible compared to other technological costs of the plant. A country with these advanced nuclear technologies will be able to generate energy economically, even from the uranium resources of sea water or granite and nuclear energy will become practically limitless. Therefore, international cooperation in technology will have an entirely different importance in the field of nuclear energy, and this leads to a unique relationship in international cooperation in the nuclear industry. The group with the most advanced leading technology, combined with the merits of large scale production and manufacture, will inevitably occupy an oligopolistic position in the world market. There can only be a limited number of enrichment or reprocessing plants to meet the world requirement for the time being, and even for nuclear power plants the manufacturers will be limited by increases in unit capacity for optimal economic scale and by the cost of development of advanced technology.

This leads to the conclusion that there should be a new scheme of industrial cooperation on an international scale to meet the new requirements of efficient development of the nuclear industry on the one hand and a competitive environment in the spirit of anti-monopoly laws on the other.

Now it seems to me that universities can play an important role in the new era of international cooperation and in the sound development of nuclear technology. Universities are traditionally free from political constraints and independent of the interests of limited industrial groups. As a place for education and fundamental research in nuclear engineering, we can contribute to the advancement of technology and the promotion of cooperation not constrained by political and limited industrial interests.

I hope this seminar will contribute to enhanced international cooperation among universities as a core for much broader cooperation in the international nuclear community so that we may tackle the problems confronting us in developing peaceful uses of nuclear energy in the coming decades for the betterment of mankind.

Discussion: Part I

Dr. SUZUKI[*]

The remarks made by Dr. Davis and Prof. Oshima emphasize the importance of international cooperation, on which I totally agree with them.

In democratic societies, I think we have some aspects in which each country has to follow public opinion or domestic preference even if it is not internationally acceptable. Let me give an example, this afternoon, the result of the presidential election in the United States will become known. In the short term, the direction of United States nuclear policy may be influenced by the result of the election. In the long term, however, it seems to me that there is a kind of social preference which, independently of the President's intention, is an underlying basis for the determination of nuclear power policy in the United States, and I myself am very much interested in the future of that preference.

While the future is certainly unknown to all of us, we seem to be seeing the possibility that, due to political reasons or others, nuclear power will be limited to domestic markets, in spite of the great cry for the need for international cooperation. In this regard, I would like to ask Dr. Davis, as a representative of the United States nuclear industry, whether or not nuclear energy would be still very attractive as a business even in the case of such a limited market.

Dr. DAVIS

Bechtel planning has been and continues to be based on the expectation of resumption of active nuclear power business in the United States and overseas at some fairly early date. It would be influenced somewhat by the election, but is the only feasible alternative within a very few years. We also are confident and planning for the development and eventual use, when needed, of breeders.

Dr. PETIT

What are the United States and the Japanese positions about possible deeper involvement of governments and central political authorities in nuclear organization and construction in the future?

Dr. DAVIS

We do not expect any shift in ownership of nuclear power plants, despite some recent utility executives' statements. One problem is financing, which must and will be solved for any utility investments. With respect to the fuel cycle the answer is not clear. Whether reprocessing, which was contemplated as private, does or

---

* Associate Professor, Department of Nuclear Engineering, University of Tokyo, Japan.

does not stay that way will likely depend on the results of the election, at least in part. Enrichment, now government-owned, will probably stay as such. We, Bechtel, tried to establish a private enrichment plant, but lost the vote in Congress. This will probably be just as well in view of subsequent events, and again may be affected by the election.

Dr. OSHIMA

It is true that there will be more increased central government intervention in the nuclear industry, especially in parts of the nuclear fuel cycle such as enrichment and reprocessing, because of safeguards against proliferation and monopoly of enterprises. However, in Japan there is a strong feeling that the market mechanism and commercial pricing should be maintained in the nuclear industry and that therefore it is not desirable to have nationalized enterprises. Maybe it is different in the United States and France.

Dr. KADOYA*

In connection with the first question raised by Prof. Suzuki about the presidential election which affects nuclear policy, I would like to know at least which candidate the United States nuclear industry is favoring. May I ask you, Dr. Davis, in short, an embarrassing question: Which presidential candidate did you vote for?

Dr. DAVIS

The Bechtel group of companies is perfectly neutral. As a member of Reagan's Energy Policy Task Force, I assume it is obvious who I voted for before leaving the United States.

Dr. KADOYA

Thank you, Dr. Davis. I just wanted to take this opportunity to express my sincere sympathy with the people working, struggling in the nuclear industry in the United States at this hard time.

---

* *Director, Ebara Corp., Tokyo, Japan.*

Part II

Toward an Acceptable Fuel Cycle Scheme

The Nuclear Fuel Cycle: An Overview

Manson BENEDICT

1. Introduction

My purpose today is to give a brief overview of the technology of the nuclear fuel cycle. I shall describe briefly the processes which are now being used and shall venture some opinions about future trends.

To keep within the time allocated me and to concentrate on the fuel cycle likely to be used in most nuclear power systems, I'm limiting my talk to various forms of the uranium-plutonium fuel cycle, those which use uranium-235 or plutonium as fissile material and uranium-238 as fertile. I don't mean to suggest that fuel cycles using thorium and uranium-233 are not important, but power systems using these fuels are farther off in time and are used in fewer systems.

2. Uranium-Plutonium Fuel Cycle

Figure 1 illustrates the uranium-plutonium fuel cycle. In this figure each rectangle, numbered from 1 to 20, represents a fuel-cycle step or process; each line, lettered from A to Z, represents a material being transferred or processed. The two rectangles with heavy borders, 9 and 18, represent respectively converter and fast breeder reactors, in which electricity is generated, the objective of these fuel cycles.

Processes 1 through 10, in the first two rows, are the steps in the once-through, slightly-enriched uranium fuel cycle to which nuclear power systems in the United States have been presently limited by Presidential directive. In these steps materials handled range from uranium at A to irradiated fuel J. In countries such as France, England, and Japan, in which reprocessing is practical and recycle of plutonium to thermal reactors is considered, the fuel cycle includes steps 11 through 15 of the second and third rows and deals with additional materials L through Q, ending with high-level waste Q in permanent storage 15. For the fast breeder reactor

Massachusetts Institute of Technology, U.S.A.

18 being introduced in France and the U.S.S.R. and under consideration in England, Germany, and Japan, the fuel cycle includes process steps
16 to 20 and materials R to Z of the last row.

3. Uranium Resources and Their Energy Potential

Uranium ore A is found in almost all parts of the world and occurs in
a great variety of minerals. At the present price of uranium concentrates
B, around $30 per pound of U_3O_8 or $80 per kilogram of uranium, ore
containing from 0.1 to 0.2% of uranium is of commercial grade.

The top part of Table 1 gives the OECD's[1] figures for the reasonably
assured uranium resources which could be produced for less than $80/kg
U and the annual uranium production in 1978 of the principal uranium-
producing countries. To compare these figures with uranium requirements
of nuclear power plants it may be noted that the annual uranium consumption of a one-gigawatt (1,000-megawatt) pressurized water reactor running
at 70% capacity factor requires about 23.8 metric tons per year of uranium
enriched to 3.3%.

To produce this in an enrichment plant rejecting depleted uranium containing 0.2% U-235 requires about 147 metric tons of natural uranium.
This involves steps 1 through 10 of Figure 1. If irradiated fuel is reprocessed and the recovered uranium and plutonium are recycled to pres-

Table 1.   Uranium resource and production data[2*] (Metric tons U).

	Resources reasonably assured at cost under $ 80/kgU	Annual production in 1978
United States	531,000	14,200
Australia	290,000	516
Union of South Africa	247,000	3,960
Canada	215,000	6,803
Niger	160,000	2,060
Namibia	117,000	2,697
France	40,000	2,183
Others*	250,000	1,491
Total, all countries*	1,890,000	33,900
Reasonably assured at cost of $80–130/kgU	740,000	
Estimated additional at under $130/kgU	2,500,000	
Total*	5,030,000	

* Excluding centrally managed economies.

[2*] from OECD.[1]

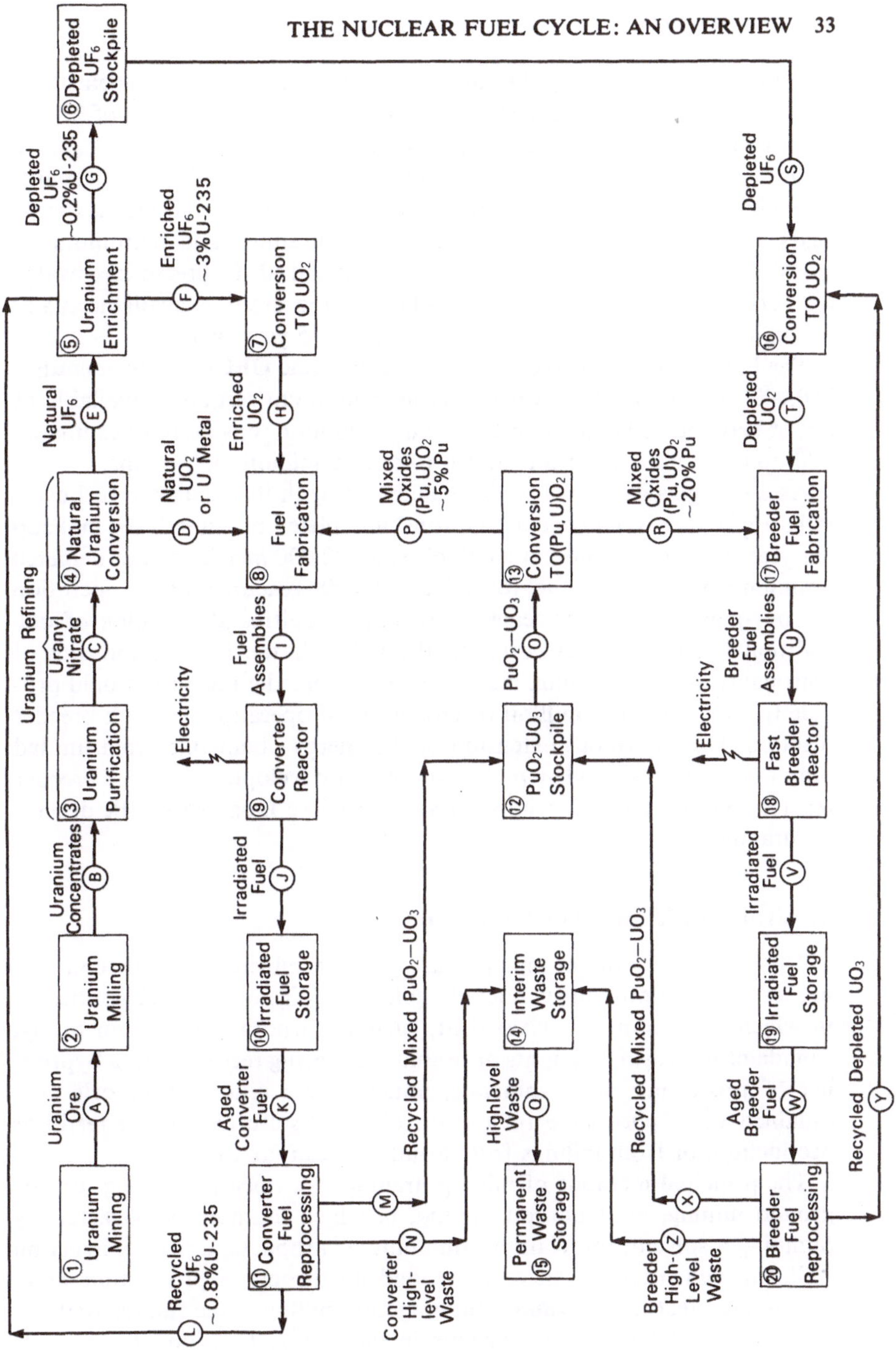

Fig. 1. Uranium-plutonium fuel cycles.

surized water reactors (steps 11 through 15 of Figure 1), the annual consumption of natural uranium is reduced to 108 metric tons, a saving of 27%.

Thus, the 1,890,000 reasonably assured uranium resources of Table 1 would support $1,890,000/147 = 12,900$ gigawatt-years of electric generation without reprocessing or recycle, or $1,890,000/108 = 17,500$ gigawatt-years with reprocessing and recycle of both uranium and plutonium. A world with 500 gigawatts of nuclear capacity, if all in pressurized water reactors, would have enough reasonably assured uranium available at $80 per kg to operate 26 years without recycle or 35 years with.

The bottom part of Table 1 shows that the same OECD report identifies 740,000 tons more of reasonably assured uranium resources available at a cost between $80 and 130 per kg and 2.5 million tons more of estimated additional resources, for a rounded total of 5 million metric tons.

As a one-gigawatt fast breeder reactor 18, with its associated fuel cycle steps 16–20, needs only one or two tons of depleted uranium S for makeup per year, the world's present stock of over 200,000 metric tons of depleted uranium would provide fuel for at least 100,000 reactor-years of operation for one-gigawatt breeder reactors, without requiring the mining of one additional ton of uranium ore. Or, the U-238 in the 1.89 million tons of reasonably assured uranium resources, in fast breeder reactors, would provide fuel for around a million reactor-years of operation for such breeder reactors. This enormous extension of the energy obtainable from limited uranium resources is what makes successful development of the breeder reactor so important, especially for a country like Japan with little domestic uranium.

4. Uranium Minerals and Mining

The principal uranium minerals are primary minerals, usually containing tetravalent uranium such as pitchblende U_3O_8, or secondary minerals in which all uranium is hexavalent, such as carnotite, potassium uranyl vanadate. Increasing amounts of uranium are being recovered as a byproduct of processing other minerals, notably from crude phosphoric acid obtained when phosphate rock is dissolved in sulfuric acid for fertilizer production, or from tailings from South African gold mines.

The principal methods of mining uranium are open-pit mining, underground mining, and solution mining, in which uranium is extracted by pumping a solvent, for example ammonium carbonate, down an injection well, through the mineralized zone, and up through a production well.

A unique feature of uranium mining and milling is the radioactivity of uranium ore. This is advantageous in facilitating location of economic ore deposits and in grading mined rock for uranium content. A price is paid for the radioactivity, however, in the precautions which must be taken to ventilate uranium mines, remove dust from mines and refineries, and prevent their refuse from spreading radioactivity. The principal hazard is from radium, its daughter the gas radon, and their decay products. As these are removed during uranium refining and form only slowly in refined uranium owing to the long half-life of their parent, thorium-230, refined uranium concentrates are much less hazardous than uranium ore.

5. Uranium Milling

In milling uranium ores[2] no one method is universally applicable, because of the great variety of uranium minerals and host rock. Uranium minerals are not susceptible to flotation and are usually too finely divided for density separation. Consequently, uranium is usually extracted by chemical leaching. The leaching agent used depends on the nature of the uranium mineral and the host rock. When the rock is a silicate or some other material insoluble in acid, sulfuric acid leaching is preferred, because it costs less and dissolves uranium values faster than sodium carbonate. However, when the rock is limestone or other material soluble in acid, leaching with sodium or ammonium carbonate is preferred. With tetravalent uranium, an oxidant suck as air or sodium chlorate must also be used.

Uranium may be recovered from leach solutions by precipitation with sodium hydroxide, by ion exhange, or by solvent extraction. Precipitation now is used only with carbonate leaching, because from acid solution too many impurities are precipitated with uranium.

Most ion-exchange processes use anion-exchange resins to effect selective separation of uranium, which forms complex sulfate or carbonate anions, from other metallic impurities, which do not. A typical anion-exchange resin for uranium extraction is a trimethylamino-substituted co-polymer of styrene and divinyl benzene (Figure 2). Types of ion-exchange equipment

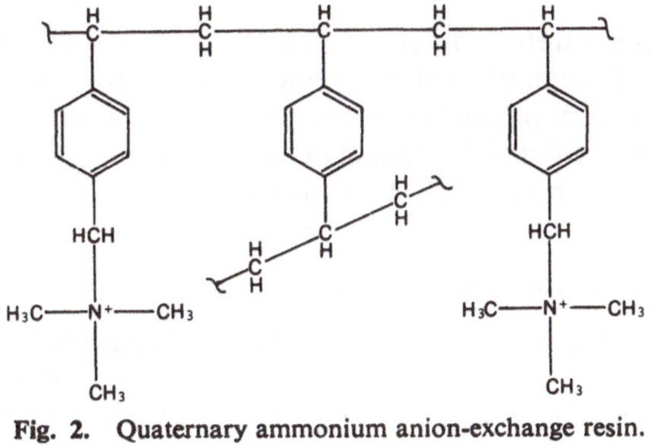

Fig. 2. Quaternary ammonium anion-exchange resin.

used include fixed-bed (in which resin is fixed in place and solutions are shifted from one contactor to another), moving bed (in which solution flow is fixed and resin is shifted from one bed to another), and continuous (in which solution and resin are alternately contacted and separated with countercurrent flow between stages).

Two highly selective classes of solvent extraction processes have been developed for recovering uranium from sulfuric acid leach liquors: the Dapex process, using di(2-ethylhexyl) phosphoric acid, and the Amex process, using trioctylamine. In both, the distribution coefficient for uranium is high, even when complexed with sulfate ion. The trioctylamine solvent is now generally preferred because it is more selective for uranium. Most new U.S. mills use this process. Uranium is stripped with concentrated sodium or ammonium chloride or sulfate.

Uranium concentrated by ion exchange or solvent extraction is precipitated, usually with ammonia as ammonium di-uranate, and constitutes the "yellow cake" of commerce.

This discussion of uranium concentration would not be complete without mentioning recovery of uranium from sea water. Although the uranium concentration is only 3.34 milligrams per cubic meter, the oceans of the world contain around 4 billion tons of uranium, which anyone with a pipe can "mine." The most promising process thus far developed for extracting this uranium selectively is ion exchange on hydrated titanium oxide. Early work in England and at Oak Ridge developed plant designs for which uranium production costs were several hundred dollars per pound. I understand that further work aimed at lowering costs is going on in Japan. Principal problems are those of handling enormous volume of sea water, preventing fouling or loss of the absorbent, and minimizing consumption of regenerant.

6. Uranium Purification

Uranium concentrates still are too impure for nuclear use. The standard method of purification (step 3) is to dissolve the yellow cake in nitric acid and separate it from impurities by countercurrent solvent extraction with a 30% solution of tributyl phosphate (TBP) in dodecane, from which uranyl nitrate C, is stripped with dilute nitric acid.

7. Natural Uranium Conversion

If natural uranium is to be used as reactor fuel, it must be converted (step 4) into metal or $UO_2(D)$. If it is to be enriched in uranium-235, today's enrichment processes require that it be converted into UF_6 (E), the most

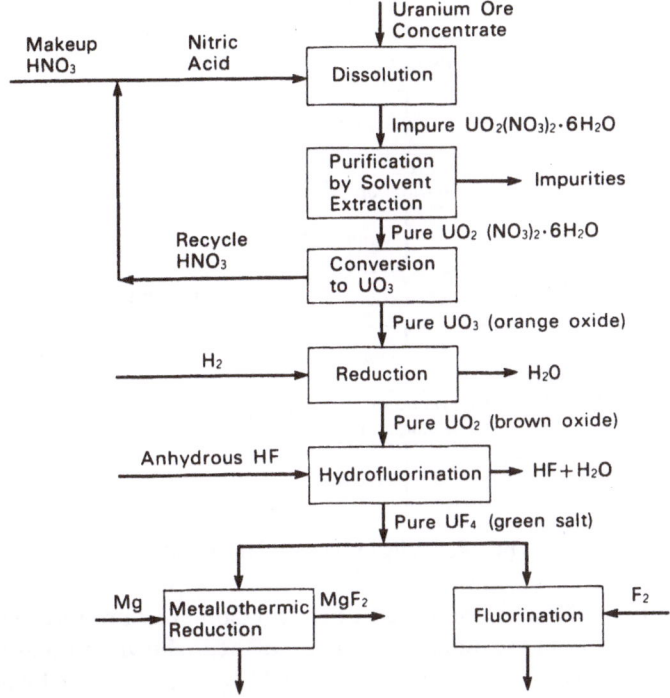

Fig. 3. Steps in conventional uranium refining processes.

stable, volatile compound of uranium. Figure 3 shows the steps in these conversion operations. Uranyl nitrate $UO_2(NO_3)_2.6H_2O$ is first converted to UO_3 either directly (in the United States), by heating to 400 °C, or in two steps (in France) by precipitation with ammonia as the diuranate follow by decomposition with steam at 400 °C. The UO_3 is next reduced to UO_2 with cracked ammonia gas at 590 °C. If the UO_2 is to be converted to metal or UF_6, it is then converted to UF_4 by reaction with anhydrous hydrogen flouride at 500 °C.

To produce metal, a mixture of UF_4 and magnesium metal in a steel vessel lined with calcium oxide is preheated to around 400 °C, at which the reduction reaction takes place with sufficient heat production to melt both products, uranium metal and magnesium fluoride.

To produce UF_6, powdered UF_4 is burned with fluorine gas in a reactor with monel walls held at 500 °C. UF_6 is purified by distillation at a pressure slightly above its triple-point pressure of 1.5 atmospheres.

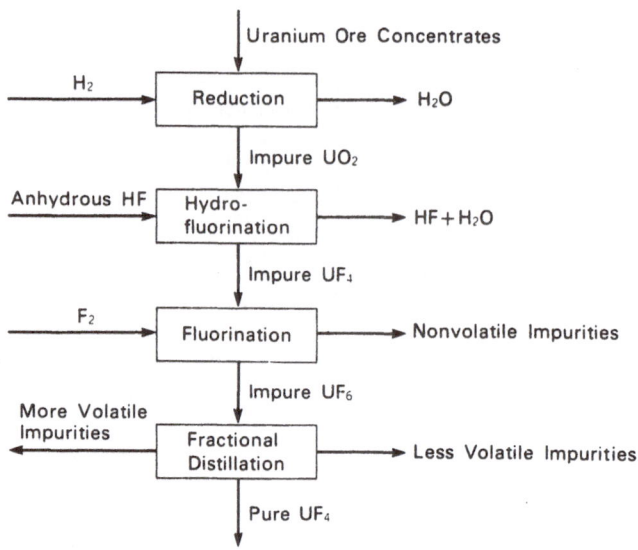

Fig. 4.   Allied Chemical UF$_6$ process.

In the United States, the Allied Chemical Company produces UF$_6$ from ore concentrates in the different sequence of steps shown in Figure 4. Concentrates are reduced directly to impure UO$_2$ which is first hydrofluorinated to impure UF$_4$ and then fluorinated to impure UF$_6$, which is finally purified by distillation. Conversion cost is around $5/kgU.

8   Uranium Enrichment

Provision of uranium enrichment services, step 5, is one of the major technical challenges of the nuclear fuel cycle. The one-gigawatt pressurized water reactor we have been using as an example uses 23.8 metric tons per year of uranium enriched to 3.3% U-235. To produce this in a separation plant depleting U-235 to 0.2% requires the production of 194,000 separative work units per year, presently costing about $100 per SWU, at a total annual cost of $19.4 million dollars, about the same as the annual cost of the natural uranium from which the enriched uranium is produced.

Table 2 lists processes for enriching uranium now in use or under development.

8.1   Gaseous Diffusion

The gaseous diffusion process has produced almost all of the enriched

Table 2.   Uranium enrichment processes.

Process	Status	Countries involved	Separ. factor	Energy use, kWh/SWU
Gaseous diffusion	Major industrial use	U.S.A., England France, U.S.S.R., China	1,004	2,500
Gas centrifuge	Large pilot plants, Major industrial use in 1980s	England, Holland, W. Germany, U.S.A., Japan	1.1	135
Separation nozzle	Large pilot plant being built	W. Germany, Brazil	1,015	3,300
UCOR	Large pilot plant being built	South Africa	1,025	3,300
Chemical exchange	Pilot plant operating	France	1,002	600
Ion exchange	Lab. development	Japan		
Atomic vapor laser	Lab. development	U.S.A.	High	Low
Molecular laser	Lab. development	U.S.A.	High	Low
Plasma separation	Lab. development	U.S.A.	High	Low

uranium used for nuclear power production. Its low separation factor requires the use of 1,200 stages to enrich uranium to 3.3%. Its high energy use contributes around \$50 to the cost of a separative work unit. Nevertheless, the process is still economically competitive with other processes. Its reliability is evidenced by the 99% onstream efficiency of the diffusion plants of the U.S. Department of Energy. When the present improvement program is complete, these plants will have a capacity of 27 million SWU/ yr and can thus serve 225 one-gigawatt nuclear power plants. The diffusion plant being built at Tricastin, France, owned by the international consortium Eurodif, will have a capacity of 10 million units per year in 1982. The U.S.S.R. plant is rumored to have a capacity around 10 million SWU/yr, of which 3 million is available for export. The English diffusion plant is small and is being phased out in favor of the centrifuge. Little is known about the plant in China.

Figure 5 is a schematic diagram of three stages of a U.S. gaseous diffusion plant. Each stage consists of a converter, which is fitted with thousands of tubes of porous diffusion barrier, a compressor to pump the process gas, UF_6, between stages, and a cooler to remove the heat of compression. When UF_6 flows through the barrier tubes, the U-235 to U-238 ratio is increased by about 0.4%. Figure 6 shows the equipment for the larger stages of a U.S. diffusion plant. The large tanks are the converters, with compressors and coolers at the rear. This stage has a capacity of about 5,000 SWU/yr. Figure 7 is an aerial view of one of the three U.S. diffusion plants, this one at Portsmouth, Ohio.

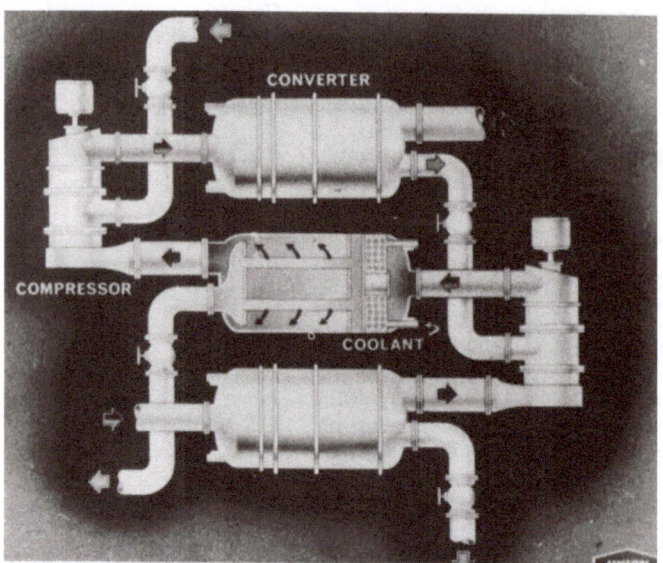

Fig. 5. Arrangement of gaseous diffusion stages.
(Photo courtesy of U.S. Energy Research and Development Administration)

8.2 Gas Centrifuge

The gas centrifuge has been chosen by the British-Dutch-German Uren-co-Centec organization as the process for its enrichment plants. At present, two plants, each producing 200,000 SWU/yr are operating, one in England, the other in Holland. By the early 1980s these plants will have a capacity of 2 million SWUs per year, with the possibility of expansion to 10 million. The U.S. is building a 2.2 million SWU/yr plant, with the possibility of expansion to 8.8 million. This process has also been chosen for use in Japan, with a projected capacity of 250,000 SWU/yr by 1985 and one to two million by 1990.

Figure 8 is a schematic diagram of a gas centrifuge. It consists of a rotor made of material with high strength-to-density ratio, such as aluminum alloy, rotating inside of an evacuated casing. UF_6 gas in the rotor is subjected to centrifugal acceleration thousands of times greater than gravity. This causes the U-235 to U-238 abundance ratio at the axis to be as much as 10% higher than at the rotor wall. A system of scoops and baffles induces longitudinal counterflow (down at the wall, up near the center in this figure), thus making the abundance ratio at the top as much as twice that

Fig. 6. View of converters and compressor.
(Photo courtesy of U.S. Energy Research and Development Administration)

at the bottom, in a centrifuge of sufficient length. A set of three concentric, stationary tubes at the axis provides means for admitting feed UF_6 to the midplane and withdrawing light fraction from the top and heavy fraction from the bottom. The high enrichment, however, is coupled with slow circulation rate and low separative capacity. Urenco machines are rumored to have a capacity of around 5 SWUs per year, Japanese machines 10, and the longer U.S. machines perhaps ten times as high, but still small compared with a gaseous diffusion stage. Thus, tens or hundreds of thousands of machines are needed for a full-scale enrichment plant.

Some of the many centrifuges in the Urenco pilot plant at Almelo, Holland, are shown in Figure 9. Figure 10 is a similar photo of a U.S. Department of Energy centrifuge pilot plant.

Because separation in a centrifuge is a thermodynamically reversible process, energy consumption is much less than in irreversible gaseous diffusion. Most of the energy is used to overcome mechanical friction and viscous losses in UF_6. In U.S. centrifuge plants the energy consumption per separative work unit is only about 5% that of gaseous diffusion. However, the capital cost of the centrifuge plant is higher. At today's

Fig. 7.   Gaseous diffusion plant at Portsmouth, Ohio.
(Photo courtesy of U.S. Energy Research and Development Administration)

price for electricity the cost of separative work from the two processes is about the same, $100/SWU.

8.3   Separation Nozzle Process

A German engineer, Dr. E. W. Becker, has developed the so-called separation nozzle process for separating uranium isotopes. The separating element for this process consists of a long semicircular groove about a tenth of a millimeter in radius, shown in transverse section in Figure 11. Feed gas, a mixture of 5% UF_6 and 95% hydrogen, flows from a pressure of about one atmosphere into a low-pressure region through a curved slit with first a convergent, then a divergent cross-section. This accelerates the gas to supersonic speed, and the curved groove downstream of the slit produces a high centrifugal acceleration. This sets up an isotopic enrichment gradient, with gas farther from the wall enriched in U-235. A knife-edge downstream from the slit divides the stream, with the more deflected portion enriched in U-235. A separation factor around 1.015 is obtained. Although this is much higher than in gaseous diffusion, the dilution of UF_6 with 19 times its volume of hydrogen gives the nozzle process about the same specific energy consumption as gaseous diffusion.

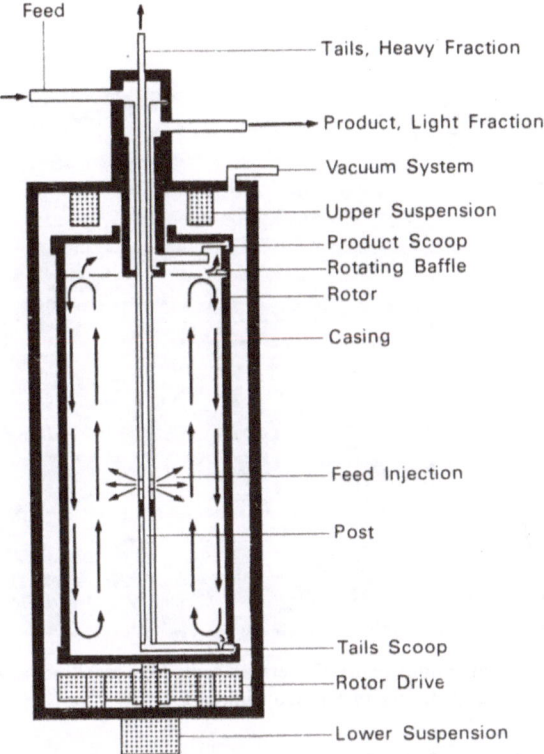

Fig. 8. Countercurrent gas centrifuge with internal circulation.

This process is used in a 180,000 SWU/yr pilot plant being built in Brazil.

8.4 South African Process

The UCOR process being developed by the Uranium Enrichment Corporation of South Africa bears some resemblance to the separation nozzle process, in that the separating element is characterized as a fixed-wall centrifuge, and the process fluid is a mixture of hydrogen and UF_6. However, there must be substantial differences, as the UCOR process operates at pressures of several atmospheres, only one-twentieth of the feed gas to a stage is taken for the enriched fraction, and the separation factor is higher. Energy consumption is about the same. Details of the separating element have not been described. At last word, a large pilot plant using this process was being built in South Africa.

Fig. 9. Urenco-Centec pilot plant of German centrifuge machines at Almelo, Netherlands. (Photo courtesy of Urenco Limited)

8.5 French Chemex Process

A promising new process for enriching uranium has been under development in France for over ten years, but few details have been disclosed. A recent paper by Dr. Coates stated that the process involves chemical exchange between unspecified uranium compounds in two immiscible liquid phases, one organic, the other aqueous.[2] Refluxing means were not disclosed. The separation factor exceeded 1,002. Contactors were pulse columns 1m in diameter by 20 m high. Power consumption would be under 600kWh/SWU. Cost would be competitive with other processes. The long equilibrium time of 15 months to make 3% enriched uranium is seen as an advantage because it precludes practical use of the process to make highly enriched uranium.

8.6 Advanced Processes

The U.S. Department of Energy has under development three advanced isotope separation processes: the Atomic Vapor Laser Isotope Separation Process (AVLIS), the Molecular Laser Isotope Separation Process (MLIS), and the Plasma Separation Process. None has yet reached the pilot plant

Fig. 10. U.S. gas centrifuge pilot plant. (Photo courtesy of U.S. Energy Research and Development Administration)

stage, but all are judged possible competitors for the gas centrifuge, certainly with higher separation factors and likely with lower costs.

The AVLIS process has been under development by Lawrence Livermore Laboratory and Jersey-Nuclear Avco-Isotopes, Inc. Figure 12 shows a form of the process proposed by the latter. Uranium metal in a water-cooled crucible is struck by a focused sheet of electrons, which heat a line of metal to 3,000 K. Uranium vapor atoms diverge radially upward from the line source and flow between cooled product-collector plates so oriented that uranium metal atoms move past them. The space between the plates is illuminated by a pulsed laser whose light is at a frequency in the visible tuned to excite U-235 atoms but not U-238. A following pulse of ultraviolet light from a second laser imparts sufficient energy to the excited U-235 atoms to ionize them, while leaving U-238 un-ionized. A magnetic field perpendicular to the plane of the figure deflects the ionized uranium atoms into the collector plates. The principal problems of this process are de-

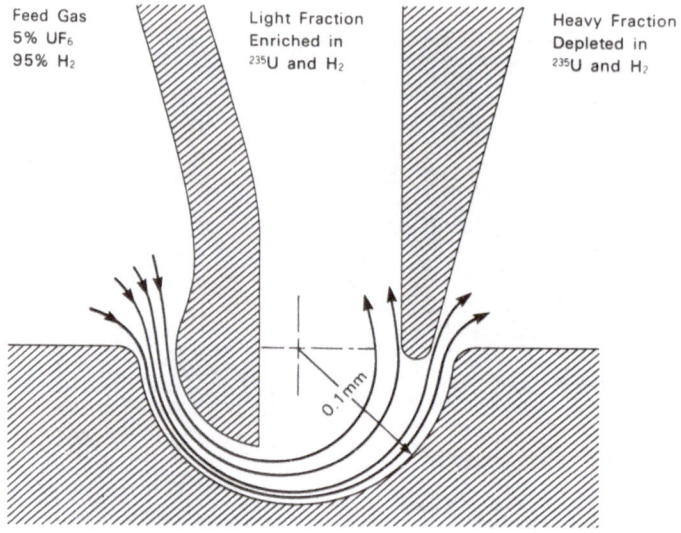

Feed Gas
5% UF₆
95% H₂

Light Fraction
Enriched in
²³⁵U and H₂

Heavy Fraction
Depleted in
²³⁵U and H₂

0.1 mm

Fig. 11.   Cross-section of slit used in separation nozzle process.

velopment of lasers of the requisite energy, repetition rate, and endurance, and handling uranium metal at high temperatures.

The Molecular Laser Isotope Separation Process, under development at Los Alamos, uses UF_6 vapor as working fluid. To obtain sufficiently selective absorption by $^{235}UF_6$, it is necessary to cool the vapor to around 75K to bring most of the UF_6 into its lowest vibrational state. Since the vapor pressure of UF_6 is effectively zero at this temperature, this calls for special measures to delay condensation till after light is absorbed. The proposal is to circulate a mixture of UF_6 and hydrogen through a hypersonic nozzle to obtain the desired low temperature after expansion. Before the UF_6 has time to nucleate and condense, the mixture is irradiated, first with a pulse of infrared light from a laser tuned to excite $^{235}UF_6$ but not $^{238}UF_6$, then with other light sources of sufficient energy to dissociate excited $^{235}UF_6$ while leaving $^{238}UF_6$ undissociated. The lower-fluoride dissociation product of $^{235}UF_6$ can then be separated from undissociated $^{238}UF_6$ by conventional means. The difficulties of this process are again the need to develop special lasers plus the problems of working at low temperature and high gas velocities.

Time does not permit discussion of the other processes listed in Table 1.

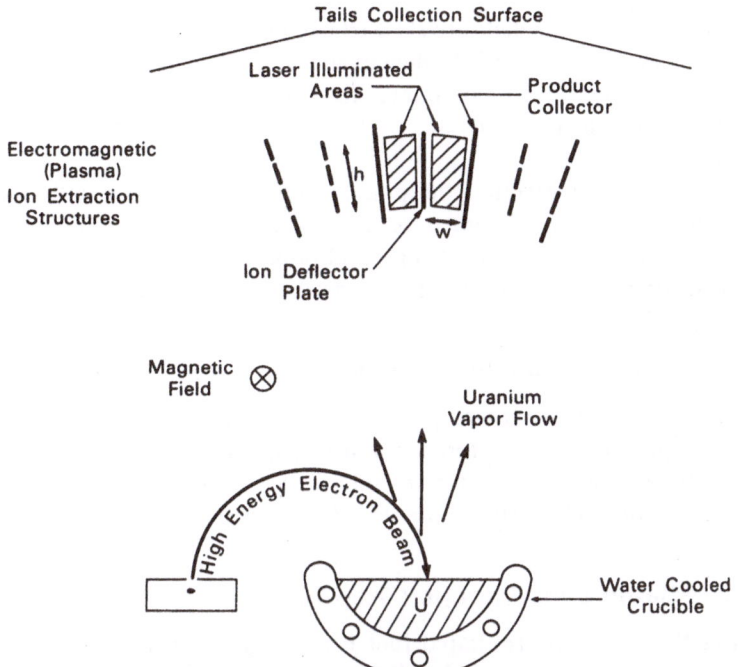

Fig. 12. Uranium metal vapor laser isotope separation process.

9. Conversion of UF$_6$ to UO$_2$

Since all enriched uranium now produced is in the form of UF$_6$, it is necessary to convert this to UO$_2$ (step 7) before it can be used as reactor fuel. One conversion method is to reduce UF$_6$ to UF$_4$ in the gas phase with hydrogen, after which the UF$_4$ is converted to UO$_2$ by reaction with steam at 650° C.

10. Fuel Fabrication

To fabricate fuel for PWRs (step 8), the process used at the Springfields Works of British Nuclear Fuels Limited[3] is representative and will be described briefly. UO$_2$ powder is milled with an organic solvent and binder. The slurry is spray-dried under conditions that produce particles of the desired size and density. Particles are formed into pellets in a hydraulic press. Binder is volatilized in a furnace at 800° C, and pellets are sintered in a hydrogen atmosphere at 1,650° C. Pellets are finished to dimensions in a centerless grinder.

To make fuel pins, pellets are stacked in stainless-steel or zircaloy tubing, to which end caps are fitted. Sufficient space is left to accommodate fission product gases. The assembled pin is filled with helium, after which the end caps are seal-welded. Welds are proved tight by mass spectrometer leak testing.

To make fuel assemblies, the pins are fastened together with spacers, end-fittings, or both, designed to maintain the correct alignment and clearance under reactor operating conditions.

Fabrication costs in the United States are around $100 per kg U.

11. Irradiation in Converter Reactor (Step 9)

In pressurized water reactors today fuel typically sustains a burn-up of 30,000 megawatt-days per ton and remains in the reactor for about three years. The uranium-235 content of fuel then is around 0.8%, and the fuel contains around 0.9% plutonium and 3.5% fission products.

12. Irradiated Fuel Storage (Step 10)

After this burn-up, irradiated fuel J no longer contributes effectively to the nuclear chain reaction and is discharged to water-cooled storage basins lined with stainless steel. The fuel is intensely radioactive and generates considerable heat, though at a declining rate. Heat production rate per ton of fuel is around 20 kW after 150 days and 10 kW after a year. Continuity of reliable cooling is essential. The water is kept clean by filtration and ion exchange and is monitored for radioactivity. If an assembly leaking fission products is detected, it is encased in a leak-tight overpack. The foregoing is the procedure adopted in the United States for storing spent fuel until decisions are made regarding more permanent arrangements.

13. Reprocessing

In other countries, where the fuel value of the uranium and plutonium in spent fuel is given more weight and where greater urgency is felt for packaging its radioactivity in a form more suitable for permanent storage, irradiated fuel K, aged for from a half-year to one or more years, is reprocessed, step 11. The principal reprocessing plants now operating are those at Marcoule and La Hague in France, Windscale in England, and Tokai-Mura in Japan, of which Figure 13 is a photograph. A plant to reprocess

Fig. 13. PNC reprocessing plant at Tokai Works, Japan.

5 tons of irradiated uranium per day was designed and nearly completed at Barnwell, South Carolina, in the United States, but its operation has been indefinitely deferred by Presidential directive. All these plants use the Purex process, with minor variations.

The principal steps in the Purex process are shown in Figure 14. The first step is to prepare fuel for dissolution, by cutting open the cladding. This is usually done by shearing the fuel bundle into short lengths, after removing external hardware. During decladding, radioactive xenon and krypton fission products are evolved and removed by off-gas treatment, step 16. Fuel is then dissolved in hot nitric acid (step 2) while the cladding hulls remain unattacked. Gases evolved in this step include oxides of nitrogen and fission product iodine. These are scrubbed with water to remove nitrogen oxides as nitric acid (step 15), and then also routed to off-gas treatment.

In feed preparation, step 3, the dissolver solution is diluted to bring its pH to 2.5 and plutonium is converted into its most extractable, tetravalent form by addition of NO_2.

In primary decontamination, step 4, uranium and plutonium are separated from over 99% of the fission products by solvent extraction with

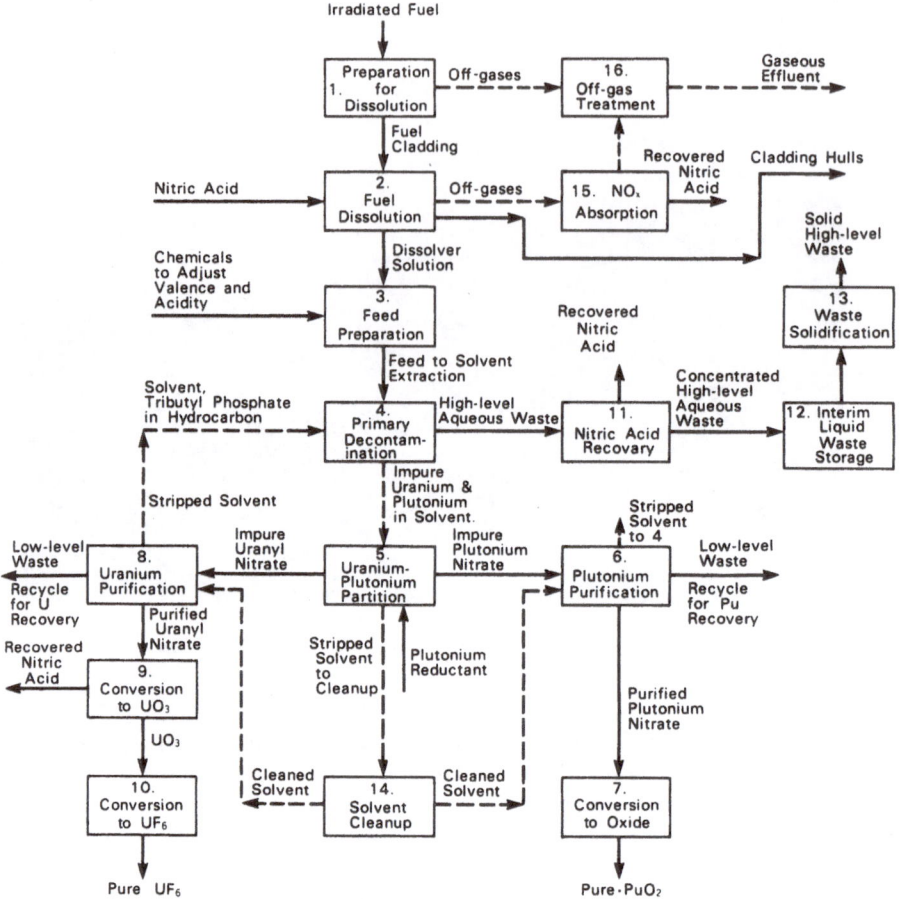

Fig. 14. Principal steps in purex process.

30 volume percent tributyl phosphate (TBP) in a paraffinic diluent. In partition, step 5, plutonium is separated from uranium by reducing plutonium to the organic-insoluble trivalent state with a reductant strong enough to reduce plutonium but not so strong as also to reduce uranium. Hydroxylamine or tetravalent uranium are preferred.

Additional cycles of solvent extraction with TBP are used to purify plutonium, step 6, and uranium, step 8. Purified plutonium nitrate is converted to oxide, step 7, either by precipitation as oxalate followed by ignition, or direct ignition. Purified uranyl nitrate is ignited to UO_3, step 9. High level wastes are concentrated by evaporation, step 11. Nitric acid recovered from these steps is recycled to the dissolver. TBP solvent, which is gradually hydrolyzed and degraded by radiation and contaminated by some fission products, is cleaned in step 14 by washing with sodium or ammonium carbonate and then recycled. Low-level wastes from steps 6 and 8 are processed for further recovery of plutonium and uranium, then concentrated for recovery of water and nitric acid.

Because of these recycle operations, nitric acid consumption is minimized and the volume of water to be discharged is reduced to an amount which can be evaporated into the effluent ventilating air flowing up the plant stack.

In off-gas treatment, step 16, iodine may be retained by adsorption on silver-coated zeolites. Processes are being developed for removing krypton and xenon by either cryogenic distillation or absorption in refrigerated fluorocarbon. The voloxidation process is being developed for converting tritium in irrdiated fuel into tritiated water in off-gases from step 1, and then absorbing the water. When these processes are fully operational, the only major radioactive wastes from a reprocessing plant should be cladding hulls and concentrated, high-level aqueous waste.

To deter misuse of plutonium, current thinking favors modifying the standard Purex process just described so that plutonium is never fully separated from uranium in the partition step and is produced mixed with uranium at a concentration no higher than needed for subsequent fuel recycle, around 5% when recycled (M,O,P, in Figure 1) in thermal reactors or 20% in fast breeder reactors (X,O,R). The Civex Process, proposed in 1978[4] is one such process.

14. Fast Breeder Fuel Cycles

The fast breeder reactor 18 (Figure 1) uses two types of fuel. The core is fueled with a mixture of 20% plutonium dioxide and 80% depleted uranium dioxide R, and the blanket uses depleted uranium dioxide T. Fuel assemblies in the core are required to sustain a burn-up of from 60,000 to 100,000 megawatt-days per ton. Hence irradiated fuel V from the core contains a higher concentration of fission products and is more radioactive than irradiated fuel J discharged from converter reactors. Irradiated fuel from the blanket contains few fission products but sufficient plutonium to provide the net breeding gain which is the objective of the fast breeder reactor. After storage, step 19, to permit fission products in core fuel to decay partially, aged core and blanket fuel W, in proportion as discharged from the reactor, is reprocessed, step 20, for recovery of uranium and plutonium X and separation of high-level waste Z.

Breeder fuel reprocessing will use the Purex process, modified as necessary by the higher concentration of fission products. It will probably be necessary to use two cycles of decontamination before separating uranium and plutonium, with the first cycle removing most of the radioactivity in contactors with short residence time, to reduce solvent degradation.

15. Waste Processing

Wastes from breeder reactors Z and converter reactors N are generally similar and will be discussed together.

When first discharged from a reprocessing plant, high-level wastes are best stored in liquid form to facilitate cooling. Preferred storage conditions are: stainless steel tanks surrounded by a secondary leakage barrier, nitric acid content between 2 and 4 molar, and temperature below 60°C, to reduce corrosion. Reliable, redundant cooling systems and a spare tank to which wastes could be transferred if another tank leaks are absolutely essential.

After the wastes have decayed such that the rate of heat generation is below a few kilowatts per ton, in from 5 to 10 years they can be adequately cooled in solid form and should be converted to a water-insoluble solid. Of the numerous solid waste forms on which work has been done, borosilicate glass is the form favored in most countries. Extensive pilot-plant work on this waste form has been done in the United States, Japan, England, Germany, and France, and a full-scale glass production plant is operating in France.

Conversion of liquid wastes to glass involves three steps: removal of water, calcination of nitrates to oxides, and conversion of oxides to glass. These steps may be carried out either in sequence or concurrently. They may be carried out continuously or batch-wise.

The French AVM process shown schematically in Figure 15 is a continuous process for converting waste sequentially, first to calcine and then to glass.[5] Liquid waste is fed to the top end of an inclined, slowly rotating, heated tube. The waste is dried in the upper half and calcined in the lower half, heated to 400°C. Calcine and glass-making solids (frit) are fed to a ceramic melting crucible heated to 1,150°C by induction heaters. Molten glass builds up in the crucible for 8 hours and then is cast into a stainless-steel waste storage canister. The canister is then disconnected, cooled, and welded shut. This process has been operated on an industrial scale at Marcoule since 1977.

A different type of glass-making furnace, which has produced up to 20 kg of glass per hour at the U.S. Pacific Northwest Laboratory, is shown in Figure 16[6]. This may be fed with frit and calcine, as in this figure, or directly with frit and liquid waste. The melter is a ceramic-lined cavity in which the glass is heated by electric current passing between immersed electrodes.

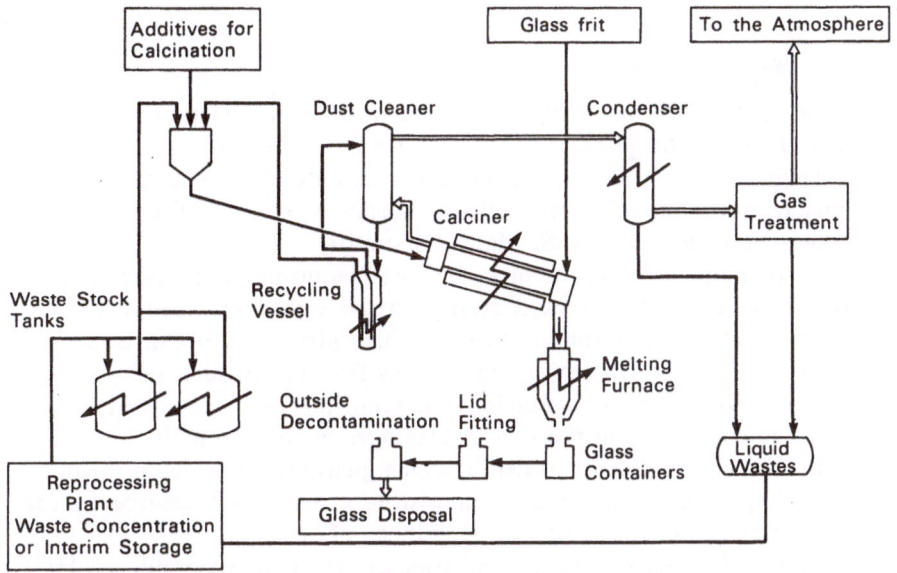

Fig. 15. The continuous process employed in the Marcoule vitrification plant (AVM Process).

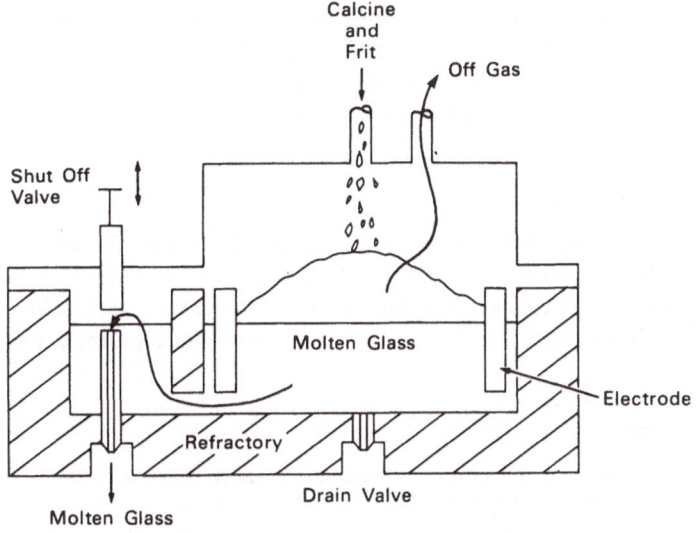

Fig. 16. Joule-heated ceramic melter.

Molten glass overflows continuously through a valve and bottom-discharge port. I will show an example of borosilicate glass made by duPont in a somewhat similar process.

16. Waste Storage

The final step in managing radioactive waste is provision of safe, long-term storage. The prime requirement here is prevention of escape of wastes into air or water contacting humans. Safe interim storage can be provided in monitored water-cooled basins, as practiced in France, or air-cooled vaults, as at the U.S. Idaho Laboratory.

For permanent storage a procedure not requiring continuous surveillance is preferred. Irretrievable storage in stable geologic formations, 500 to 1,000 meters underground, through which groundwater can be shown not to circulate, is the storage mode now favored. Bedded salt deposits have been identified as a suitable waste repository in West Germany and are being considered in the U.S. Advantages of salt are its high thermal conductivity, its plasticity at depth, which guarantees that fissures will be self-sealing, and the absence of circulating groundwater assured by the stable existence of the salt beds for millions of years.

Granite, favored in Sweden and Canada, is another suitable geologic formation, when demonstrably free of circulating groundwater.

To provide additional assurance against escape of radioactivity from wastes in underground storage, stainless steel waste canisters can be provided with an even more corrosion-resistant overpack. Titanium has been considered in the United States and copper in Sweden.

With all these precautions, inadvertent migration of radioactivity from an underground repository can assuredly be prevented.

17. Conclusion

In this overlong talk, I have been able only to outline the several process steps of the nuclear fuel cycle. I'll be glad to give more details by answering questions. I'd like to leave the impression that fuel cycle technology is quite well established and that the processes are feasible, economic, and safe.

References

1. Organization for Economic Cooperation and Development: "Uranium-Resources, Production and Demand," Paris (1979).
2. J. H. Coates: "France's Chemical Exchange Process," International Conference on the Nuclear Fuel Cycle, Amsterdam (1980).

3. H. Rogan: "Fuel Manufacturing Technology and Production Facilities at BNFL Springfields," British Nuclear Fuels Limited (1977).
4. M. Levenson and E. Zebroski: "A Fast Breeder System Concept: A Diversion-Resistant Fuel Cycle," 5th Energy Technology Conference, Washington (1978).
5. D. W. Clelland *et al.*: "A Review of European High-Level Waste Solidification Technology," in Proceedings of the International Symposium on the Management of Radioactive Wastes from the LWR Fuel Cycle, Report CONF–76–0701 (1976), pp. 137–165.
6. J. L. McElroy *et al.*: "Waste Solidification Technology, U.S.A.," *ibid.*, pp. 166–189.

Reprocessing Technology in Europe

Cyril BUCK

This paper will cover the outline, general philosophy, and status of European technology for the reprocessing of irradiated uranium oxide fuel elements discharged from light water reactors and AGR-type reactors. It will deal with receipt and storage of fuel, shearing and dissolution, the recovery of uranium and plutonium, the purification and concentration of uranium and plutonium before transferring them to conversion plants for the manufacture of oxides, and the pretreatment of the solid liquid gaseous waste arisings. The possible impact on reprocessing technology of the recent International Nuclear Fuel Cycle Evaluation (INFCE) will also be considered.

As a result of the cooperation between reprocessing interests in France, the Federal Republic of Germany, and the United Kingdom under the aegis of United Reprocessors, and the long experience at La Hague, Karlsruhe, and Windscale, the proposed plants in the three countries tend to follow a common pattern in their overall specifications. It is therefore possible in this paper to describe a typically European plant but call attention to special features proposed in individual plants. Table 1 shows the overall specifications for proposed plants in the three countries. It should be noted that there will be some variation in the size of the units.

1. Outline of the Process

The receipt and storage facility will be a part of the reprocessing plant complex that will provide a suitable buffer between reactor discharges, fuel shipments, and reprocessing, and will allow a further cooling period before reprocessing takes place.

Fuel elements will be transferred directly from the storage facility into the "head-end" section of the reprocessing plant, and will be checked to

Reprocessing Division, British Nuclear Fuels Limited, U.K.

Table 1. Comparison of specifications for industrial scale reprocessing plants (INFCE, 1980)

Topic	U.K.	France	Federal Republic of Germany
Commercial program	6,000 T in first 10 years	400 T/A Extension to 800 T/A mid 80	1,400 T/A
Maximum burn-up	40 GWD/T	40 GWD/T	40 GWD/T
Fuel cooling	~ 3A(annum)	1A minimum	~ 3A
Mode of operation	Continuous on shift	Continuous on shift	Continuous on shift
Process	Chop/leach solvent extraction	Chop/leach solvent extraction	Chop/leach solvent extraction
Design capacity	5 T/D	—	4 T/D
Plant availability	250 D/A	250 – 300 D/A	300 D/A

ensure that fuel cooled for no less than one year is fed to the plant. The fuel elements will be fed into a shear machine, and the resulting short pieces of fuel pins, about one to two inches in length, will pass forward for dissolution in nitric acid. Off-gases from this operation will be segregated. An absorption and retention system for iodine will be fitted, and as a minimum the plant will be designed so that krypton-85 removal can be back fitted when suitable processes have been developed. After dissolution of the fuel the section of fuel pins known as "hulls" will be monitored for the presence of residual fuel and passed to the conditioning unit for storage and eventual disposal. The dissolver liquors will pass forward through a clarification stage and thence to the accountancy tanks before entering the separation section.

The separation system will follow the well-known Purex-type process using tributyl phosphate/odorless kerosene as the solvent.

Co-decontamination and partitioning of uranium and plutonium will be performed in the first extraction cycle. Uranium and plutonium will be subsequently purified in two separate lines of two extraction cycles each. U(IV) will be used as the reductant for plutonium. High-active, segregated medium-active, and low-active liquid wastes will be concentrated before passing to the appropriate conditioning process for long-term storage or disposal. Liquid wastes will be:

highly active waste concentrate (HAWC) ($> 10^4$ Ci/m^3);
medium-level aqueous waste concentrate (MAW) (10^{-1} to 10^4 Ci/m^3);
organic liquid waste;
low-active liquid waste (LAW) (10^{-1} Ci/m^3).
Solid wastes will be:

cladding and structural material of fuel elements, residue of feed clarification;

ion-exchange resins and iodine absorber material;

off-gas and exhaust air filters;

Pu-contaminated waste, decontamination material, and engineering wastes.

In addition there will be gaseous wastes which will require treatment before discharge.

2. Discussion of Technologies

2.1 Spent Fuel Reception and Storage

Irradiated assemblies will arrive in approved transport flasks by sea, rail, or possibly road; normal civil engineering techniques will be adequate for the construction of such entry facilities.

For plants that are to be constructed within the next decade, storage of fuel and the associated reception and discharge facilities will probably be carried out in ponds. The storage of fuel in water ponds is a proven technology. Experience exists with wet storage of LWR and HWR spent fuel for periods up to 20 years with low burn-up and for shorter periods with higher burn-up fuels. Moreover, observation and investigation work is continuing to ensure that the behavior of fuel in the reprocessing plant will not be affected by underwater storage conditions. Dry storage techniques are being developed, but at the present time the advantages of flexibility that exist with pond storage would appear to favor the latter.

Ponds can be economically built in modules of 1,000 T fuel capacity. The fuel handling capacity should be adequate to meet the requirement for quick turnaround of transport flasks and associated transport equipment. The fuel will be stored in fixed geometrical array to avoid criticality problems. Neutron-absorbing material can be incorporated in the racks to increase storage density, but this must depend on the overall economics. The ponds will be provided with water cooling and water clean-up facilities. In the case of the U.K. fuel will be stored in multi-element bottles to ensure pond and building cleanliness. There is some divergence of view on the degree of protection necessary to meet the possibility of damage from aircraft crashing on this section of the plant.

2.2 Chop and Leach

Fuel can be transferred directly within the shielding from the storage pond to the chop and leach section of the reprocessing plant. Considerable experience has been gained in the chopping of irradiated fuel in shearing machines, and present designs cover special provisions for feed arrangements, remote maintenance of the shear machine, blade replacement, and safety features. Similarly, experience has been gained on the special requirements of the dissolution process such as fuel washing and chemical techniques using soluble poisons such as gadolinium, which are necessary for the safe operation of this equipment. The off-gases from this section of the plant require special treatment. Equipment for the removal of iodine, carbon-14, and nitrous fumes is available. Whereas on the scale of reprocessing operations in the past, the removal of krypton-85 was not important, methods are now under development for its retention. Plants will be designed to segregate the krypton-85-bearing streams, and arrangements will be made to direct them to a suitable concentration and storage unit when it is considered that a satisfactory plant can be constructed.

Clarification of the dissolver liquor will be necessary. A suitable centrifuge has been specifically designed for this purpose and it is undergoing trials under operating conditions with special attention to the handling of solid residues.

2.3   Separation and Purification

The equipment used for the separation of fission products and the purification of uranium and plutonium has reached an advanced stage of development, and in many cases there is an impressive history of successful operation. The essential features required are short residence time in the early sections of the solvent extraction process, safety against criticality incidents, and complete reliability under the extreme testing conditions of reprocessing operations. The alternatives of pulsed columns or centrifugal contactors can be considered for the first cycle separation and plutonium purification sections, and mixer settler or centrifugal contactors for the uranium purification section. The methods for measuring the flow of active liquids and their transfer between vessels are all well tested in operating plants, as are sampling, in-plant instrumentation and measurement, and analytical techniques. Special instruments for particular applications will require final proving.

2.4   Uranium Product Stream

Multi-effect evaporators are proven equipment for the concentration of this stream, and there is a long history of successful operation of thermal denitration plants for conversion to oxide, the normal end product for the reprocessing plant.

2.5   Plutonium Product Stream

A suitable evaporator for concentration of this solution has been in service for many years. Equipment for precipitation and conversion to plutonium oxide is available while equipment for the co-precipitation and co-conversion to mixed plutonium and uranium oxide is under development. The choice will depend upon the final product required.

2.6 Liquid Wastes

2.6.1 Highly Active Liquid Waste Streams

HAW wastes after concentration can be stored in specially designed high-integrity storage tanks. There is a long experience of successful operation. However, it is recognized that these wastes should be solidified at the earliest practicable time. Throughout the world a very large engineering and scientific effort has been directed towards glassification of waste, and the AVM (Atelier de Vitoification Marcoule) process is now working successfully. The glass product can be safely stored awaiting the final disposal step when a means acceptable to the general public has been agreed upon.

2.6.2 Medium-Active Liquid Waste Streams

There are well proven single and double effect evaporators available for concentration of medium-active wastes. These concentrates will be immobilized by incorporation into matrices, a commonly favored form being bitumen, for which stage established processes and equipment are available. Interim storage will be provided until an acceptable disposal route is agreed upon.

2.6.3 Low-Active Liquid Waste Streams

Again, concentration and incorporation into concrete or bitumen is a well-established technology, and disposal routes are agreed.

2.6.4 Degraded Solvent

The solvent will be treated for extraction of plutonium, uranium, and neptunium and stored awaiting the development of disposal techniques,

2.7 Solid Wastes

2.7.1 Cladding

After thorough washing, the cladding can be compacted mechanically and packaged or set in a matrix of metal, plastic, or concrete for storage.

2.7.2 Resins and Sludges

Processes have been demonstrated for:

—the incorporation of resin and sludges into a cement matrix with production of cement blocks for storage and disposal;

—incorporation into bitumen and containment in steel drums;

—incorporation into a plastic matrix.

2.7.3 Contaminated Equipment

The process will involve cutting and baling to reduce the volume of waste followed by the application of surface coatings prior to packaging for disposal.

2.7.4 Plutonium-Contaminated Solid Waste

Incineration is well developed for these materials followed by acid leaching for recovery purposes. Materials of furnace construction may still present problems in operation and maintenance. Packaging methods are under development.

3. Philosophy of Design

In general European technology follows closely the philosophy of design accepted for the base case plant by INFCE.

3.1 Provision of Capacity

Although the reprocessing plant is mainly assumed to run on a single line process, some provision for redundancy needs to be made in the layout of the plant. This redundancy will depend upon the detailed maintenance philosophy adopted and the commercial risk the operator is prepared to take. For example, it has been the practice to duplicate the highly active first-cycle separation section of the plant because decontamination prior to repair could be comparatively lengthy.

3.2 Maintenance Philosophy

The generally accepted maintenance philosophy comprises three different systems, each with its own application to the appropriate section of the plant:

  (i) Remote maintenance by installed mechanical devices;
 (ii) Modular replacement involving decontamination removal and replacement of components;
(iii) Contact maintenance which involves decontamination entry to the cells and repair or replacement of equipment by adaptation of normal engineering methods.

Remote maintenance will be employed for mechnical items in the highly active parts of the plant such as the shear cell, which will be equipped with remotely operated in-cell hoists, manipulators, specially designed tools, and decontamination facilities.

The modular replacements will be applied chiefly to the items whose life is expected to be shorter than that of the whole plant. Items such as evaporators and calandrias will be designed for partial decontamination and removal by means of hoists and replacement using remotely operated joints or welding.

Contact maintenance will be applied to the highly active and medium-active solvent extraction plant and similar areas. In certain cases its use may presuppose a standby line of similar equipment in a separate cell if maximum use is to be made of the commercial potential of the plant. However, contact maintenance is a high-cost item in plant operation, and the need should be minimized by the careful choice of equipment and detailed design. In all cases, very high standards of fabrication and construction will be required.

Plant items will, as far as possible, be fabricated from all-welded corrosion-resistant stainless steel or other alloys specially chosen to resist corrosion. The use of moving parts inside the process cells will be minimized, *e.g.*, by employing vacuum lifts for liquor transfers.

Equipment for the supply of services or inactive chemicals will be located outside the biological shielding and will be maintained directly.

Remote maintenance techniques will be rehearsed on full-scale mockups, thus providing highly trained personnel to carry out the operations with maximum efficiency.

3.3 Safety Philosophy

Once the general layout of the reprocessing plant has been determined, detailed analysis can be performed to determine the specific safety requirements. Items of general importance are radiation protection, containment of radioactivity, and criticality control.

3.3.1 Radiation Protection

The plant will use well-established technology for radiation protection measures. It will be built according to internationally accepted standards for shielding and operated within the guidelines of environmental radiological safety recommended by the International Commission on Radiological Protection (ICRP).

As a practical note, it is becoming accepted practice to design plant shielding so that operator dosage is limited to 0.5 rem/man/year in carrying out normal duties, with a control limit of 1.0 to 1.5 rem/man/year to take account of other special duties.

3.3.2 Containment

Because of the special properties of the material, a multiple containment system will be applied in those sections of the plant where fissile and radiating material is handled. Precautions will be as follows:

(i) Radioactive material will be treated or stored in a series of independent enclosures: stainless steel process vessels; heavily shielded process cells; an outer building with appropriate protection.

(ii) To reduce potentially harmful events to a minimum the following precautions will be taken: minimization of nuclear material transfer between installations; avoidance of high pressure and elevated temperature in the process; precautions against interaction of process.

(iii) Based on postulated modes of failure including those due to external hazard, the damage arising from the following events will remain within predetermined limits: activity release; interaction between different components; radiation exposure to operators and public.

3.3.3   Ventilation Philosophy

The aerial discharge will comprise air from operating areas, the cells, and the vessel ventilation system.

Prior to discharge to the environment, exhaust air from cells will pass through high-efficiency filters. Vessel ventilation will be pretreated, passing through scrubbers and electrostatic precipitators before being filtered. The filters will be designed to reduce the entrained activity to acceptable levels and will contain all forseeable incidents in the plant.

According to their radioactive material release potential, the sections of the building will be connected to differently graded negative pressure zones to prevent the spread of contamination within the building. The pressure gradient ensures a directional flow from rooms with a lower contamination potential to those with a higher radioactive inventory. The lowest pressure will be maintained in the cell block.

3.3.4   Criticality Control

During the fuel processing, the fissile material may exist in solid form, but more generally it will be in the form of aqueous or organic nitrate solutions. Measures will be taken to prevent criticality excursions. The aim of all methods is to assure subcritical conditions under all circumstances.

Important parameters affecting criticality are the following: initial enrichment of fuel; burn-up; mass concentration; moderation; presence of neutron poisons; geometry; reflection and interaction.

The control of criticality will be exercised by the following measures: administrative measures such as mass limitation or concentration limitation checked by analysis and/or on-line monitoring, *e.g.*, with neutron monitors; application of soluble or heterogeneous neutron poisons; geometrical limitations of the equipment in connection with precautions against neutron interaction.

The transition from nuclearly favorable to non-controlled equipment will be protected by the use of on-line instrumentation, analytical control, and, where necessary, physical devices.

4. Environmental Impact

There is little doubt that this topic could be the subject of a separate and specialist lecture to an appropriate symposium. For the present it is sufficient to note the study carried out in INFCE Working Group 4 where the ICRP dose limitation system was reviewed. It was concluded that the radiological impact of both reprocessing and subsequent recycling operations may be carried out safely within the ICRP recommendations. The report went on to say that up to the year 1995, on the basis of a total world reprocessing capacity of 10,000 t/annum, reprocessing operations as a whole may increase the global general public exposure by about 0.025 to 0.1 %—*i.e.*, well within the margin of regional and temporal fluctuations expected from natural background radiations.

Another INFCE group concluded that the estimated contribution of waste disposal to the collective dose commitment is small when compared with that from natural background irradiation, and of the same order as that from other stages of the fuel cycle combined.

5. Minimization of Proliferation Risk

This topic has assumed great importance in the international nuclear community, and INFCE had as one of its principal tasks to consider effective measures that could be taken in the fuel cycle to minimize the danger of proliferation of nuclear weapons without jeopardizing energy supplies. The problem centered on the management of plutonium once it has been inevitably produced by nuclear power plants. INFCE Working Group 4 therefore arrived at definitions of possible routes to proliferation with respect to reprocessing plants, assessed the risks of this occurring, considered the sensitive parts of the reprocessing and recycle system, considered the technical options open to reduce the risks of proliferation, and concluded that safeguard measures will be of greater importance than the technical measures reviewed and should be regarded as of similar importance to those for safety and physical protection. The possible effects of non-proliferation measures on existing reprocessing technology are discussed below.

5.1 Definitions
For the purpose of their evaluation, the INFCE Working Group:
  (i) Referred to the misuse by a government of nuclear fuel cycle facilities, know-how, or materials as *proliferation;*
  (ii) Referred to attempts by sub-national groups to acquire nuclear materials from fuel cycle facilities as *theft;*
  (iii) Defined *diversion* as those activities needed to implement a decision to misuse facilities or attempt manufacture of nuclear weapons. Diversion may be either "overt" or "covert".

5.2 Overall Assessment and Sensitivity
After discharge from the reactor and during periods of comparatively

short cooling, the plutonium in fuel elements is protected by a barrier of high irradiation, and this creates technical barriers that are difficult for individuals or sub-national groups to overcome. Similarly, during the early stages of the reprocessing plant, the high irradiation and heavy shielding provided to protect operators also provides barriers against diversion. However, such barriers to a large extent disappear in the later stages of the reprocessing cycle and these areas can be regarded as sensitive and requiring consideration, as also can the transport and storage of plutonium.

6. Technical Measures Aimed at Increasing Diversion Resistance

6.1 Co-processing

The application of co-processing to produce a mixed plutonium/uranium nitrate solution provides a further reduction in vulnerability against theft and overt diversion. Plutonium nitrate no longer exists in the fuel cycle. Its effectiveness could depend upon the ease of modification of the plant. The process is technically feasible, but process control problems have been encountered with a co-processing flow sheet. Differences of opinion emerged on the extent of further work required, but there was a view that it should be considered as an option for a future generation of plants.

6.2 Co-conversion

The plutonium nitrate recovered during reprocessing is first blended with uranyl nitrate and co-converted into a mixed oxide. Pure plutonium oxide no longer exists in the fuel cycle. The present technology of oxalate precipitation is unsuitable for co-conversion but a number of possible alternative methods are being studied, and development work would be required for the design of an industrial plant.

6.3 Storage and Transport of Plutonium as MOX

Following the process of conversion, the oxide powder of uranium and plutonium can be physically blended. Some protection in transport and storage would be provided. However, no experience of long-term storage of MOX exists, and technical problems in fuel fabrication can be foreseen.

6.4 Co-location of Plant

Co-location of reprocessing plants, plutonium stores, MOX fuel fabrication, and waste treatment plants was seen as a desirable long-term objective and likely to develop as a normal arrangement.

6.5 Additional Radiation Barriers to Protect Plutonium

Consideration was given to partial reprocessing, spiking of fuel, and pre-irradiation of fuel. Partial reprocessing was regarded as *not* effective, since for the foreseeable future, out-of-pile recycle times are likely to be such that protection would not be provided. Considerable development of technology would be involved. Spiking of fuel and pre-irradiation of fuel were not supported on either technical or non-proliferation grounds.

6.6 Physical Barriers

The physical isolation of reprocessing plants and downstream facilities by means of structural barriers may be expected to improve safeguard-ability significantly and decrease opportunities for diversion. This feature will be mentioned in the following section on *safeguards*.

INFCE concluded that while the technical measures mentioned earlier have an influence on reducing the risk of theft, they have a limited influence on reducing the risk of proliferation. It was judged that safeguard measures are more important than technical fixes.

7. Safeguards

The purpose of *international safeguards* as operated by the appropriate international organization is to detect and deter diversion, and their effectiveness depends closely on the probability of detection. Clearly the quantity of materials that might have been removed and the timeliness of detection are important. The safeguard systems have been built up on the basis of verification of nuclear material accountancy supplemented by containment and surveillance.

At the scale of plants now in operation, these measures are considered to provide effective safeguards. However, it will be difficult in the case of future *industrial reprocessing plants* for material accountancy to meet the limits of accuracy required because the quantities processed will be so large. Detection times will also pose problems. The reprocessing plant will work on a continuous basis, and this means that the plant cannot be stopped and cleaned out frequently to strike material balances. Thus, in future plants, much more emphasis will be placed on a combination of containment and surveillance measures appropriate to the particular situation.

The impact on future plants on the development of safeguards can be summarized as follows:

(i) Effective safeguards are an essential feature of the industry.

(ii) Safeguard requirements must be included in the initial design.

(iii) A balance must be struck between an effective safeguards system and overall plant costs.

(iv) The chemical plant and biological shielding provide two barriers to unrestrained access to nuclear materials. In present technology these

barriers must be breached for the addition of reagents and maintenance of the plant.

(v) There is considerable room for the extension of the use of in-pro-
cess material accounting methods. The development and demonstration of these techniques will be followed, and their effectiveness
considered.

Two proposals have been made in Europe for improvement in containment and surveillance.

(a) One concept for the future is that the whole operational area should
be contained within a substantial building to which access through a minimum number of entrances is monitored by the International Inspectorate. Wastes (where measurement of nuclear material is difficult) must
be transferred to storage along contained routes.

(b) A concept known as Pipex (with perhaps a longer time scale) is that
heavy shielding would be extended to all stages of the process, including
conversion and later stages, and that there should be a minimum number of
outlet points with guaranteed inacessibility to plutonium at all other points.
The International Inspectorate would survey the integrity of this outer
barrier, and the balance across the plant would be obtained solely by quantitative control at the inlet for irradiated fuel and at the outlet for plutonium products. Effluents will be treated within the containment to remove
or recycle nuclear material. Equipment will be decontaminated and monitored before removal through the barrier.

8. Summary of Present Position

This is necessarily an outline paper only, but it can be concluded that as
far as European technology is concerned the main chemical flow sheet can
be regarded as proven, and tested equipment is available for the main
process units. The technology is a result of gradual evolution over a period
of 30 years.

In Europe, through the machinery of official inquiries and hearings, the
governments of France and the U.K. have made the decision that plants of
the type described in this paper will be constructed safely, economically,
and in the public interest. In Germany, due mainly to the problem of public
acceptability, construction has been deferred.

The technology exists for the conditioning and safe storage of the wastes
produced from a reprocessing plant. There remains the problem of obtaining public acceptance for the form and method of ultimate disposal of these
wastes.

What further remains to be done?

Current development programs for the optimization of the operational flow sheets will be completed. Development effort of this type must be considered as an ongoing requirement in order to improve the product and improve the overall efficiency and costs. Similarly current equipment development programs aimed at proving detailed operational techniques and improving operating efficiency will be completed.

For any particular plant detailed safety and environmental impact statements will be prepared for submission to and consideration by the national authorizing body. This exercise is one of the major tasks in a reprocessing plant project and requires a large resource allocation for both preparation by the project group and assessment by the national authorizing body.

For any particular plant the engineering organization must be built up for site planning and development, final equipment choice, and layout and detailed design of the plant. On this latter point, it is stressed that because of the difficulties of maintenance and modification the success of a reprocessing plant depends heavily upon the excellence of its detailed design. Fabrication capacity of the requisite high standard must be established, and a construction organization also capable of meeting very high standards must be set up.

The recent study by INFCE on reprocessing plants should be considered since this reflects the views and concern of the international nuclear community on the operation of such plants. In the INFCE report there is complete support for international safeguards, and certainly for the next generation of plants there will be a need to provide improved safeguard systems to meet the new requirements of the higher throughput plants. This will have an impact on overall philosophy of design and plant costs.

I wish to acknowledge the work of my colleagues in INFCE and BNFL on whose work I have drawn in this lecture.

Management of the Nuclear Fuel Cycle

Ryohei KIYOSE

The nuclear fuel cycle, as a whole, is an extremely complicated large system composed of a great number of various treatment and separation processes based upon the highly developed new technologies; it should be investigated very carefully and rationally from the operational and managerial viewpoints.

1. In-core Fuel Management

The in-core fuel management, which includes core burn-up optimization and fuel reloading optimization, can be considered a sub-system of the total system of nuclear fuel cycle, and the sub-system is comparatively well defined and can be formulated as an optimization problem. In other words, the burn-up optimization is the control of power distribution in the core to minimize the fuel cost for one core life under the constraints of required power generation and maximum power and temperature limits. Optimization of the fuel reloading program is the selection of reload fuel fraction and its loading pattern to minimize the fuel costs for the total plant life, considering also the transition from the initial fuel loading to the equilibrium fuel loading. For these optimization problems, several methods of mathematical programming, such as the maximum principle, dynamic programming, and linear programming, can be applied. However, for practical application of these theoretical methods, some modification of the burn-up calculation model by comparison with measurement data of power and burn-up distributions are needed. For these kinds of burn-up and fuel reloading optimizations, automated and computerized core performance monitoring systems have been developed and are in practical use. As an example of burn-up optimization, a simplified flow diagram of a computer code OPROD for automatic generation of optimal control rod

Department of Nuclear Engineering, University of Tokyo, Tokyo, Japan.

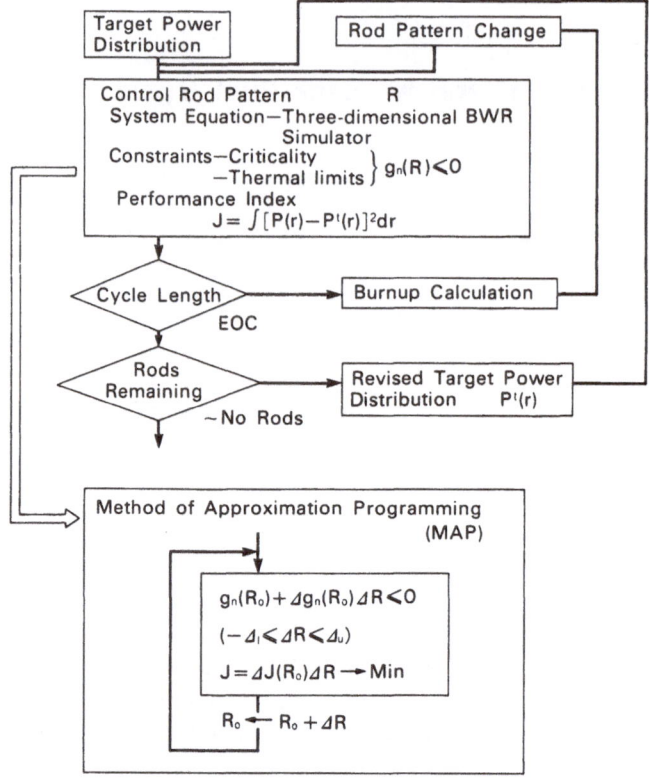

Fig. 1. Simplified flow diagram of OPROD.[1]

programming for BWRs is shown in Figure 1.[1] The entire in-core fuel management program will be as shown in Figure 2.[2]

2. Upstream Fuel Management

As uranium resources in Japan are scarce, only about 10,000 short tons, it is very crucial to maintain a stable supply from abroad of the uranium source material needed for the nation's nuclear power program. Japanese utility companies have made long-term contracts with Canada, United Kingdom, Australia, etc., to import about 157,000 short tons, and they also have shares of about 20,000 short tons in the international cooperative enterprise in uranium mining at Akouta in Niger. The total amount of about 177,000 tons of natural uranium is considered to be able to supply nuclear power reactors in Japan until about 1990. Expecting an increase in demand for uranium, the Power Reactor and Nuclear Fuel

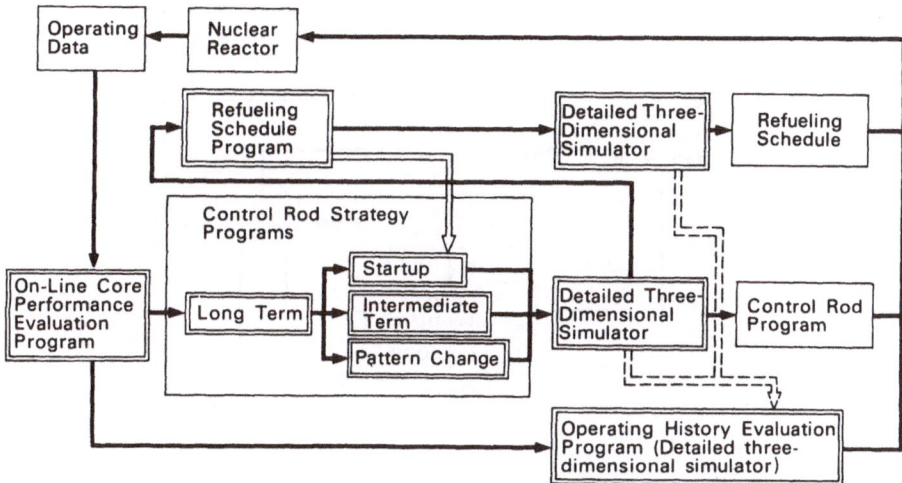

Fig. 2. BWR core management programs.[2]

Development Corporation (PNC), as well as nine private enterprises, are working in cooperation with foreign groups for exploration of uranium mines in Africa, Canada, Australia, and so on. In particular, the PNC has sent about 260 engineers to Mali and spent about ¥1.5 billion on their exploration project in fiscal 1980. Technological development of the direct production of UF_4 by a wet process (PNC process), which can be connected with the heap leaching of uranium ore and with the UF_4-UF_6 conversion process, has also been carried out by the PNC. A pilot plant of 200 tons U/yr for UF_6 production is now under construction.

Recovery of uranium from sea water, where uranium exists in very low concentration (about 3 ppb) but in very large amounts (about 4 billion tons), is very attractive for Japan. A conceptual design of a pumping-fixed bed system has been made, as shown in Figure 3 and Table 1. Another method of sea current direct utilization has also been studied.[3]

Concerning uranium enrichment service, Japanese utility companies have already contracted with the United States Department of Energy (USDOE) for enrichment service corresponding to nuclear power capacity of about 51 GWe, and with the EURODIF company corresponding to about 9 GWe. At the same time, a domestic project of uranium enrichment by the gas centrifuge method has reached the pilot stage. For the PNC pilot plant of about 7,000 centrifuges, the first 1,000 centrifuges began operation in September 1979, and the next 3,000 centrifuges in October 1980. The

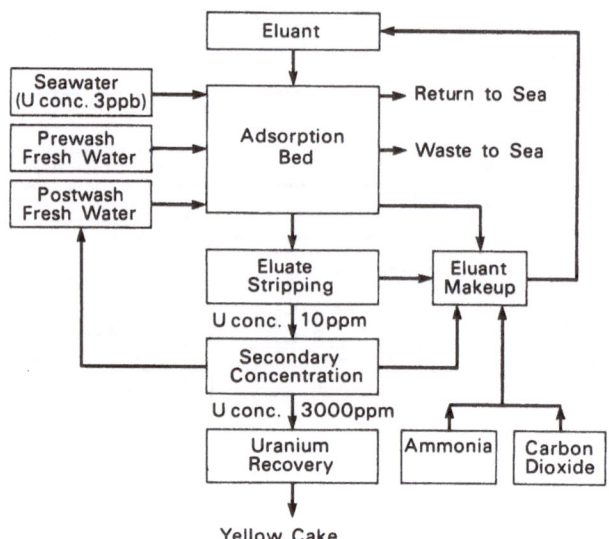

Fig. 3.   Flow diagram of "pumping-fixed bed" system with eluant of $(NH_4)_2CO_3$.[3]

Table 1.   Basic conditions of pumping-fixed bed system.[3]

1. Annual uranium production	100 t U
2. Concentration of uranium in sea water	3 ppb
3. Adsorption recovery	60%
4. Adsorption structure	Fixed bed, multi-layered
5. Adsorption-elution cycle	10 days/2 days
6. Adsorbent	$TiO_2 \cdot nH_2O$
7. Adsorption capacity	200 μg U/g·ad./10 days
8. Bulk density	1.0 g/cm³
9. Linear flow rate of sea water	60 cm/min
10. Granule size	1 mmϕ
11. Adsorbent loss	0.1%/cycle
12. Bed thickness	10 cm
13. Elution recovery	95%
14. Elution temperature	Room temperature
15. Elution solution	1 M $(NH_4)_2CO_3$, $(NaHCO_3, Na_2CO_3)$
16. Uranium concentration in eluate	10 ppm
17. Decarbonation method	Steam stripping (Electrodialysis)
18. Secondary concentration	Ion exchange (Ion flotation)
19. Concentration recovery	95%
20. Final concentration of uranium in solution	3,000 ppm

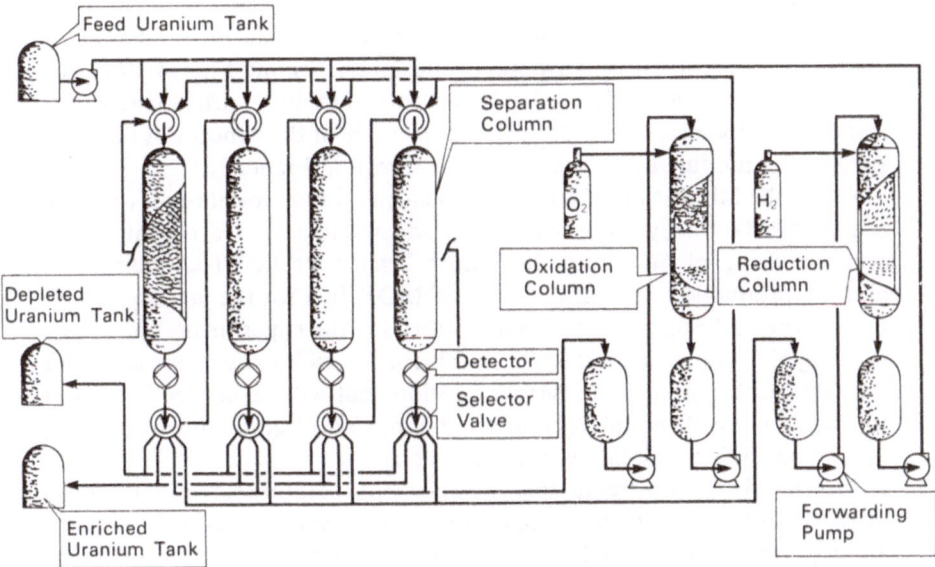

Fig. 4. Flow diagram of chemical-exchange separation unit.[4]

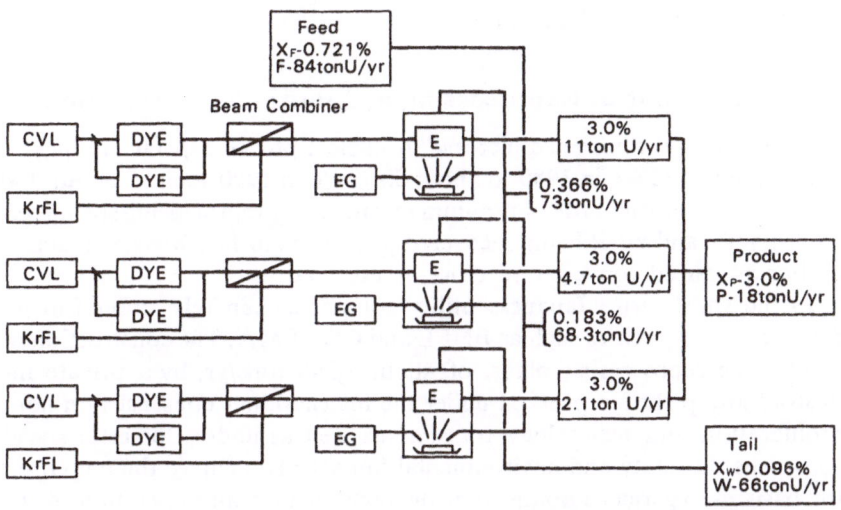

Fig. 5. Concept of a uranium enrichment module by the laser method.[5]
Note: CVL: copper vapor laser; DYE: dye laser (5915 Å and 6140 Å);
KrFL: KrF laser (exicimer laser); EG: electron gun; E: electrode.

remaining 3,000 centrifuges will be installed in the later half of 1981. A chemical-exchange process using redox chromatography in adsorption columns is also currently under development, and its flow diagram is shown in Figure 4.[4] Furthermore, at the Japan Atomic Energy Research Institute (JAERI) and other research institutes, uranium enrichment by the laser method is also being studied. An example of the concept of uranium enrichment module by a laser method is shown in Figure 5.[5]

As for the fabrication of LWR fuel elements, Japan has its own capacity of about 1,000 tons/yr at four plants in private industry. Development and manufacturing of plutonium fuels have been done by PNC since 1966, and the production of about 20 tons of MOX fuel for the prototype D_2O power reactor "Fugen" and about 1.5 tons for the experimental fast reactor "Joyo" have been completed. In fabrication technology, especially for the case of plutonium fuel, further development will be needed in respect to quality control, remote operation, gaseous and liquid effluent control, volume reduction of solid wastes, and so on.

In general, for upstream fuel management, careful technology assessment should be made from many managerial viewpoints—safety, operability, environmental effects, and proliferation resistance—and also from the economical viewpoint, especially for new methods of uranium enrichment because of the large investments required for construction and for research and development.

3. Management of Reprocessing and Plutonium Utilization

Considering the expected increase in nuclear power capacity in Japan, from about 15 GWe in 1980 to about 50 GWe in 1990 and to about 100 GWe in 2000, efficient use of uranium resources, by reprocessing spent fuels from LWRs and by utilizing recovered plutonium in fast breeder reactors and/or advanced converter reactors, is essential.

In the short term, Japanese utility companies can rely upon foreign reprocessing capacities such as BNFL and COGEMA. The construction of a full-scale reprocessing plant, of about 1,200 tons/yr, by a private industrial group should proceed under the international consensus on nonproliferation and technology transfer, as well as under domestic social consensus on safety and environmental impact. To achieve the target, on the basis of experience obtained in the construction and operation of the PNC reprocessing plant of 210 tons/yr, it will be necessary to further improve the processes themselves, to establish methodologies and criteria for safety and environmental assessment, and to develop nuclear materials management technologies based upon the principles of containment, surveillance, and material accountancy.

As to the process improvements, at the PNC plant tighter control over release of radioactive liquids has been accomplished by adding two stages of evaporation, and developmental facilities for recovery of krypton, asphalt solidification of low-level wastes, solvent recovery, denitration of uranium product solution, and co-conversion by microwave heating of the U-Pu product have been constructed. Co-processing of U and Pu at the extraction stages is also being studied using the Operation Test Laboratory. Since 1978, the Tokai Advanced Safeguards Technology Exercise (TASTEX) has been conducted with the cooperation of Japan, the United States, France, and IAEA. The R & D items of TASTEX are shown in Table 2.

Table 2.   R & D items of TASTEX.

A. Surveillance Equipment for Receiving and Storage Areas for Spent Fuels
B. Non-destructive Gamma-spectrometry of Spent Fuels
C. Hull Monitoring Systems
D. Load Cells for Weight Measurement of Stored Liquids
E. Electromanometers for Liquid Level Measurements in Adjustment Tanks
F. Applicability of DYMAC System
G. Pu Concentration Measurement by the Gamma-Ray Absorption Method
H. Pu Isotopic Composition Measurement by High Resolution Gamma Spectrometer
I. Surveillance System for Pu Handling Areas
J. Sampling Method for Resin Particles
K. Isotopic Safeguards Technology
L. Measurements of Amounts of Nuclear Materials in Adjustment Tanks by Methods of Elemental and Isotopic Correlation
M. Volume Measurements of Adjustment Tanks by Tracer Methods

Nuclear criticality safety management is especially important in design and operation of reprocessing plants. Protective methods taken for criticality safety in some principal processes at the PNC plant are listed in Table 3.

Table 3.   Criticality safety design of PNC reprocessing plant.

1. Spent Fuel Storage Pool: Space between Spent Fuel Elements
2. Dissolvers: Geometry and Concentration
3. Extraction Equipment: Geometry and Concentration
4. Adjustment and Internal Storage Tanks: Geometry and Concentration
5. Evaporators for Pu Solutions: Geometry
6. Evaporators for U Solutions: Geometry and Concentration
7. Fluidized Bed Denitrators: Geometry
8. Storage Tanks for Pu Product Solution: Geometry, Concentration, and Fixed Neutron Poison (Cd Lining)

4. Radioactive Waste Management

This problem is most important for public acceptance of nuclear power. For realistic application of the "As Low As Reasonably Achievable

(ALARA)" principle, technological developments for reduction of radioactivity releases from various nuclear facilities are needed, and research and development work on removal and/or recovery of radioactive krypton, iodine, and tririum produced in reprocessing plants are under way at various research institutes.

The accumulation in Japan of solidified low-level wastes has reached about 220,000 of 200-liter drums in 1980, of which about 80% has come from nuclear power plants and the balance from other facilities such as fuel fabrication and reprocessing plants and various research facilities. According to the nuclear power program, the accumulation is expected to increase by about 10 times in the year 2000. Both sea disposal and shallow land burial are under consideration for the final management of these low-level solid wastes. For these purposes, new solidification methods other than cement solidification, such as in asphalt and plastics, and also technologies for volume reduction by incineration, pelletization, and so on, are to be investigated.

In Table 4, some favorable properties such as better volume reduction and durability of plastic solidification products are shown in comparison with cement and asphalt solidification.[6] Leaching tests of asphalt solidification products, corrosion tests of waste containers, and measurements of migration and distribution of radionuclides in the aerated sand layer are being conducted at JAERI, corresponding either to sea dumping or to shallow land disposal.[6]

Table 4.   Properties of various wastes solidified with polyethylene, asphalt, and cement.[6]

	Polyethylene	Asphalt	Cement
Volume reduction ratio			
Granular resin	2.0	1.5	0.4
Powdered resin	1.5	1.5	—
Filter sludge	1.5	0.5	0.5
Compressive strength (kg/cm²)	200–300	Poor shape stability	180–230
Leaching fraction* (%, for one year)	0.3	—	55

* Granular resin, cesium.

Regarding high-level wastes, an accumulation of about 1,400 cubic meters of vitrified waste is expected in Japan by 2000, and R & D programs on various items as shown in Figure 6 are being conducted or planned mainly by PNC.[7] Table 5 shows the design criteria of the Engineering Test Facility which was completed recently at the Tokai Works of PNC.[7]

At JAERI, for safety evaluation of high-level wastes, a Waste Safety

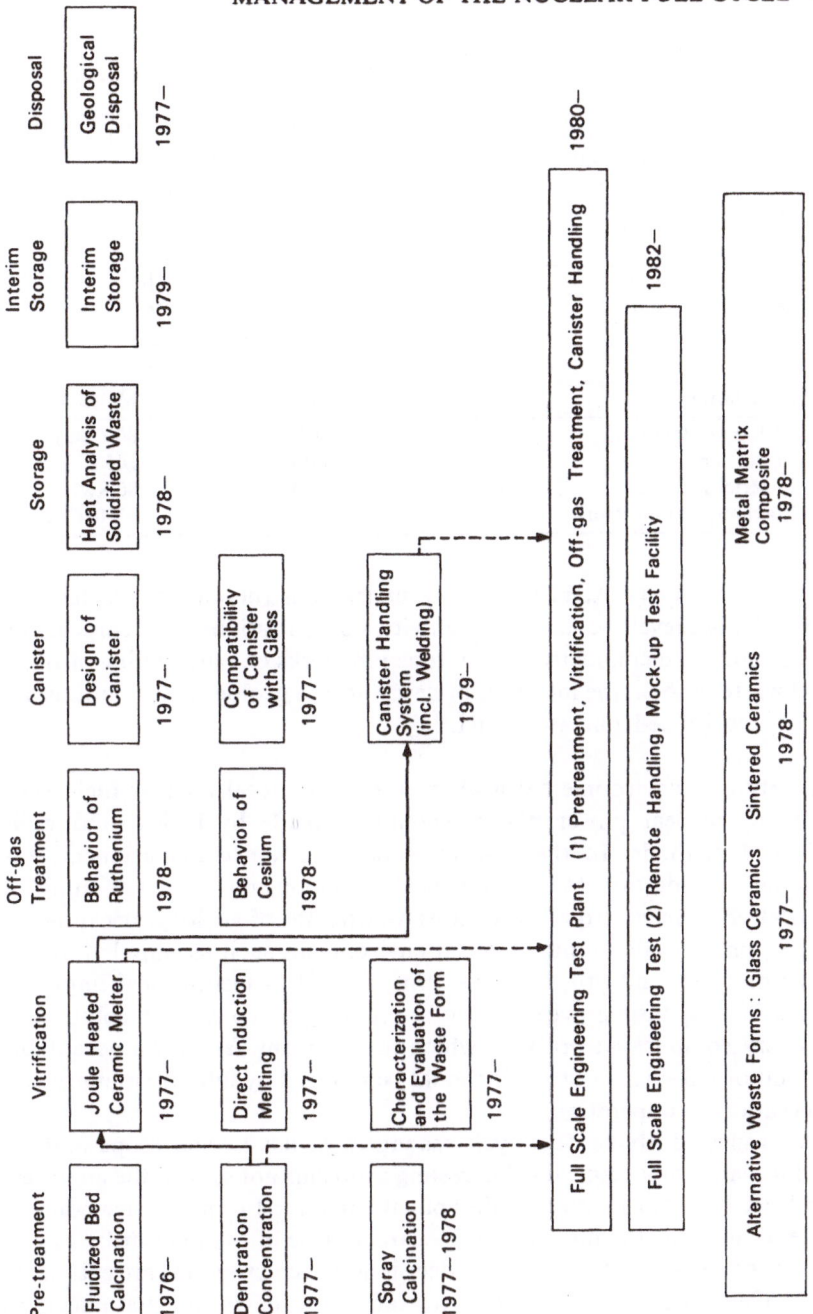

Fig. 6.   R & D programs on high-level radioactive waste management.[7]

Table 5.   Design critteria of engineering test facility.[7]

Pretreatment process		
Denitration-concentration capacity	945 /batch/day	
Denitration temperature	95°C	
Concentration	Factor 2	
Glass frit	135 kg/day	
Off-gas treatment process		
Processing capacity	50 N m³/Hr	
Materials treated	Particulates	
	Ru, NOx,	
	Water vapor	
Vitrification process	Melter B	Melter C
Melting capacity	80 l /day	80 l /day
Vessel capacity	160 l	145 l
Electricity power	100 kw	80 kw
Maximum temperature	1350°C	1350°C

Testing Facility (WASTEF) is now under construction, where tests of vitrified high-level wastes with activity concentrations of reprocessing wastes are to be conducted. Some research works on partitioning of high-level wastes and on the influence of decay heat in geological formations are also being carried out at JAERI.[6]

Safety and environmental assessments of the total nuclear fuel cycle, including nuclear power plants, should be made by logical and well-balanced methods. In the case of radioactive waste management, for example, in addition to technological developments, understanding of natural phenomena from the various viewpoints of geology, oceanology, and so on, considerations of social environment such as population distributions and land utilization patterns, as well as institutional improvements for responsible management are necessary. To reach a solution of this very complicated problem which has so many facets, an acceptable method of risk-benefit analysis should be pursued by interdisciplinary and international cooperation.

As discussed above, fuel cycle management has a wide scope and includes many important and interesting problems not only at the engineering level but also of an interdisciplinary nature. From the viewpoint of systems analysis, the most peculiar characteristic of the nuclear fuel cycle is that the risks accompanying the decisions on the programs are extremely high. These risks are concerned with safety and environmental effects as mentioned before, but also include risks of large-scale technological research and development in a more general sense. In other words, these high risks are due to the large amount of financial investment and the long time span of the projects, as well as the great number of factors to be considered in assessment of the risks.

To try to make a logical and persuasive evaluation of such a complex and large nuclear fuel cycle system seems to be too ambitious. However, with the cooperation of many young able scientists and engineers, this challenging interdisciplinary problem can hopefully be solved in the near future.

In this respect, the role of nuclear engineering departments at universities, whose responsibilities are to train those young scientists and engineers and to carry out basic research work based upon the long traditional history of engineering, will become more and more important and essential for the new technological society.

References

1. T. Hayase and H. Motoda: *Nucl. Tech.*, **48**, 91 (1980).
2. H. Motoda *et al.*: *Nucl. Tech.*, **36**, 294 (1977).
3. M. Kanno and K. Saito: 4th Joint Meeting of The Mining and Metallurgical Institute of Japan and The American Institute of Mining, Metallurgical and Petroleum Engineering, Tokyo (1980).
4. M. Seko *et al.*: *Nucl. Tech.*, **50**, 178 (1980).
5. T. Arisawa and Y. Naruse: *J. Atomic En. Soc. Japan*, **22**, 79 (1980) (in Japanese).
6. K. Araki: OECD/NEA RWMC Meeting, Paris (1979).
7. S. Suzuki: Annual Meeting of Am. Nucl. Soc., Las Vegas (1980).

Discussion: Part II

Dr. PIGFORD

It is difficult to add anything to such a comprehensive review by the previous speakers, except that I am relieved that the future for nuclear engineering will be less uncertain soon, because of the American election, than it was when we started this program this morning. There are a few more very brief comments.

One thing that is interesting to me is the interaction of the fuel cycle with the choice of reactor—for example, the thorium fuel cycle. I will try to give some brief comments. In my view, the American program is a two-reactor program (for example, water reactors and plutonium). And I think it is reasonable. We know that thorium is difficult to compete, because to introduce it purely for electrical power production, a new and different fuel cycle is too expensive.

In Canada, the situation is different because Canada is a one-reactor economy, very carefully chosen—no breeder, no fast breeder, no enrichment—and there, in the long term, the same type of reactor can utilize thorium very effectively. Perhaps that is the place for thorium and a national program. Or perhaps if you have a reactor concept that has some unique feature, like a reactor producing a very high temperature, useful for steel, and if the economics show that it is good, that is the unique place for thorium.

I agree with Dr. Buck that technical fixes really are not the way of obtaining safeguards. Now, we have learned the idea of the so-called denatured fuel cycle using thorium, where the uranium fissile isotope can be diluted with non-fissile uranium. I think that we have learned that it does not achieve very much.

With regard to reprocessing, we have heard from Dr. Davis, and he has shown that, on a financial basis, the incentive to reprocess is only a few percent if you are a country with plenty of uranium ore. That, of course, is the situation in the United States, but is not the situation in other countries, and I hope the people in the United States and other countries are beginning to recognize that difference. Reprocessing is important, and it has other features.

Professor Benedict has mentioned the importance of high-level radioactive waste management. Now, if we treat spent fuel as radioactive waste and put it in geologic isolation, we are, of course, wasting an extremely valuable resource. And, designing the geologic depository, we find that it is more difficult to provide the same kind of isolation with spent fuel for many reasons. The waste form is already specified. It contains about 50 times as much toxic transuranic radionuclides as reprocessed high-level waste. And in the United States, we are beginning to realize that human intrusion into a geologic depository is one of the risks that is important, and what better incentive is there for human intrusion than to put the valuable unreprocessed fuel there?

Now, just a few more comments on radioactive waste. We have a great problem in the United States: what criteria to use for the long-term management of radioactive waste. I would like to support Dr. Buck's recommendation that we use the ICRP (International Commission on Radiolgical Protection) dose limit criteria. I am afraid that this is very confused right now, and it may turn out that the criteria that are forthcoming in the United States will be quite a long way from criteria now used for reactor licensing. This is partly connected to the uncertain time scale. The time scale is important. Many years ago, we made calculations of the potential for contamination, compared with uranium ore. I am afraid we do not know if uranium ore could be licensed under the new criteria that are forthcoming. I think that we need to apply the new techniques that are now available to quantitative analysis of the risks from geologic disposal, and not depend upon subjective ideas. They are so hard to understand and implement.

But I do not think the high-level waste is the most difficult problem, the most exotic problem. The transuranic wastes from fabrication of recycled plutonium and from reprocessing probably will have more difficulty in meeting the same criteria as high-level waste, though they should. They contain about the same amount of transuranics. In my view, they must meet the same kind of specification. Because there are so many kinds and so much volume, I think that it is more difficult than with high-level waste. To meet the criteria forthcoming for high-level waste may be the most difficult problem.

What we call mill-tailings, the large piles of material containing radioactive thorium and radium, produced when uranium ores are purified, probably now are the most important radioactive wastes in the United States, Canada, and other countries. I believe that Canada is beginning some very good work in treatment of those, and I think that it needs to be emphasized in the future.

Dr. TAKASHIMA*

One comment I want to make here is with regard to the disposal of high-level wastes. I think we started out regarding them as something abominable. We are trying somehow to treat them and isolate them away from us. It seems to me that we have considered high-level waste disposal to be basically very defensive or very passive. But many people, including Dr. Nakamura of the PNC (Power Reactor and Nuclear Fuel Development Corporation), are suggesting something I find interesting about some elements of fission products, such as ruthenium, rhodium, and palladium. Uranium-235, when split, yields this platinum group by about 10%. Ruthenium can be sold for ¥2,000 ¥–4,000 per gram. It is surely an expensive rare metal. We might as well stock these materials for 20 or 30 years till we can recover them and perhaps use them for social benefit. So we cannot necessarily say high-level waste is dirty: we have profitably attractive metals in it. In the case of plutonium fuel, you have again such platinum group by more than 15%. I want to emphasize this point.

* Professor, The Research Laboratory for Nuclear Reactors, The Tokyo Institute of Technology, Tokyo, Japan.

At this time, I would like to address a question to Prof. Benedict on the U.S. program towards the peaceful use of nuclear energy. It started with sharing your gaseous diffusion plants in Paducah, Oak Ridge, and Portsmouth. Each of these three plants came up to more than 15,000 tons of separative work units per *annum* capacity, and the design for that capacity must have been made before 1950. So how did the U.S. arrive at that decision in 1950 to build such large-scale gaseous diffusion plants? If they were to use 90% enriched uranium solely for military use, then the U.S. would be able to produce 75 tons per *annum* of enriched uranium. That seems too excessive to me. If you had been thinking of a peaceful power use of 3% enriched uranium, then you would have come up to 4,000 tons per *annum* of product. This amount is enough to supply all the power plants in the world today. It is also a tremendous capacity. So what I want to know is what was the background against which the U.S. made that decision in 1950. It is still a question to me today.

I think our present nuclear energy scheme is mostly due to that decision about the three plants. If it had not been for that decision, and those big plants, we could have chosen, perhaps, a different nuclear energy scheme. It would have been more varied without that decision, I believe.

Dr. BENEDICT

I do not like to intrude on a conference devoted to the peaceful uses of atomic energy the true answer to the questions which have been asked here. But going back historically to 1950, there was, in the United States, a great concern about our country's military posture relative to the Soviet Union. It was thought that we needed to increase our supply of fissionable materials. And one of the two routes for obtaining more fissionable materials was to extract more uranium–235 at 90% concentration from natural uranium. So the decision was made consciously to increase the capacity to 17,000 tons of separative work units per year, by building the enlarged Oak Ridge plant, the Paducah plant, and the Portsmouth plant and operating them as a complex to produce 90% enriched uranium.

At about the same time, however, there was again an offshoot of a military development. It was found that the water-cooled pressurized water reactors, which were developed first for submarine propulsion in the United States, were adaptable economically to the production of electricity for civilian purposes. So, about 1958 or 1959, the first pressurized water reactors were built, and it was realized that their use of 3% enriched uranium fuel represented an economic option. So, in a sense, we are converting our swords into ploughshares. We are finding in the United States that these large gaseous diffusion plants, which were first built strictly for military reasons, are now being devoted almost exclusively to the production of 3% enriched fuel for powering the civilian nuclear power plants of the United States, Japan, and many other parts of the world. I think that this is a very favorable and hopeful development, and that, in a sense, we are converting our nuclear swords into ploughshares. And I hope this will partly answer your questions.

Dr. TAMIYA*

I have the privilege of being the last commentator. A number of people have already mentioned a lot of things. Professor Pigford has mentioned that there is one certainty in an uncertain future. Just before coming here I watched a TV program which mentioned that Mr. Reagan was elected, and therefore one of the uncertainties was eliminated.

Well, another thing I would like to call your attention to is the point made by Prof. Kiyose about an acceptable nuclear fuel cycle. Having listened to all the speakers today, it seems that technically nuclear fuel cycle technology is well established. In addition to that, these past three years nuclear power has been stepping backward slightly or moving slowly, due to the reactor accident at Three Mile Island and the International Nuclear Fuel Cycle Evaluation (INFCE), although the motivations were different and they were not done intentionally. It was a matter of acceptance. In the case of the reprocessing plant at Barnwell, the United States government did not reach the level of acceptance. The problem of Three Mile Island was that of public acceptance. As to the outcome of Three Mile Island, the acceptance of the public has been more or less, at least internationally, positive.

Dr. Buck has been the co-chairman of the INFCE program. We can say that the compatibility of non-proliferation and the peaceful use of atomic energy can be established. And the needs for peaceful use are certainly different in different countries, as was rightly mentioned by Prof. Pigford. So I am very happy that the Americans are beginning to understand these factors, and I believe it was one of the major achievements of recent years. There are two kinds of acceptance. One is a political acceptance on the level of non-proliferation, and this should move in time toward international consensus. At the same time, public acceptance is very important, and I certainly believe that needs for nuclear power are different in different countries due to different conditions. But, against this background, it is important to proceed with the development of nuclear power on the basis of international consensus.

* Director, Japan Nuclear Fuel Service Company, Limited, Tokyo, Japan.

Part III

Engineering Philosophy on Safety

The Management of Nuclear Safety: Lessons Learned from the Accident at Three Mile Island

Thomas H. PIGFORD

1. Introduction

The accident at Three Mile Island (TMI) revealed that to have nuclear safety there must not only be reliable equipment, but there must also be competent and qualified people. At TMI there were equipment malfunctions, but the vital safety equipment performed well. An accident with serious core damage and radioactive release occurred only because operators mistakenly stopped the flow of emergency cooling water. Health effects from radiation exposure were negligible. Continued loss of flow of emergency cooling water would have led to core melting, but the containment would have survived and protected the public. Serious fright and trauma resulted from technical errors and public announcements of these errors a few days after the accident. Earlier experience from other reactors and analyses dating back to 1972 should have alerted the industry, the regulatory agency, and the operators and avoided the accident. This experience from the Three Mile Island accident provides many valuable lessons in the management of nuclear safety. Several such lessons revealed from the investigation by the President's Commission are discussed here.

2. Lessons Learned

Lesson 1: Plant Equipment Must be Adequately Maintained

At TMI improper maintenance probably caused the turbine to stop. This may have happened because the same air supply used for actuating valves was improperly used to clean ion-exchange equipment. After the turbine stopped auxiliary feedwater should have flowed to the steam generators to cool the shutdown reactor, but the auxiliary feedwater valves were illegally closed. For many weeks prior to the accident there had been a known leak in the relief valve that opened to relieve the transient pressure but failed to close. Later symptoms, which should have alerted the operators that a real accident was occurring, were incorrectly attributed to the

Department of Nuclear Engineering, University of California, Berkeley, California, U.S.A.

leaking valve. Charcoal filters to remove radioactive iodine from air had deteriorated and were operating below design specification. Fortunately, the filters were still sufficiently effective during the accident.

Although these malfunctions were not the causes of serious core damage and radioactive releases, they did distract the operators. Management of nuclear safety must include proper maintenance so that reliable equipment is always available and so that operators can diagnose important events when they occur.

Lesson 2: Reliable Safety Equipment is Important, and It Must be Used Properly

When the minor equipment malfunctions required the actuation of emergency cooling at TMI, the safety equipment performed promptly and reliably. The relief valve, shown in Figure 1, opened automatically after the turbine trip and the reactor shut down a few seconds later. When the relief valve stayed open, some of the reactor coolant escaped into the containment, and the reactor pressure decreased. The emergency cooling water was injected automatically to cool the reactor. The malfunction of the open valve did not interfere directly with safety, because it provided a path for the heated water and steam from the emergency coolant to escape

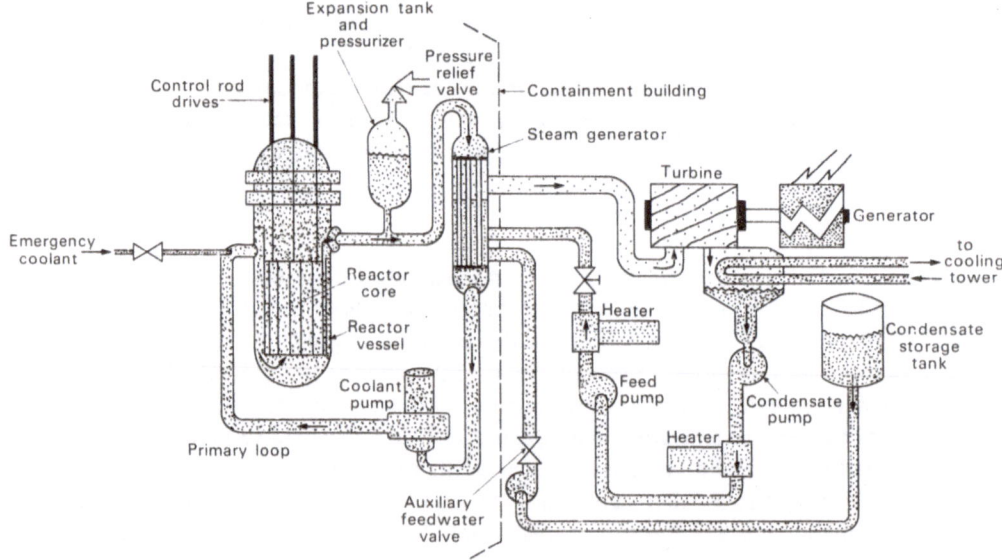

Fig. 1.   Schematic of the Three Mile Island reactor.

to the containment. The reactor was then being adequately cooled. If it had been allowed to continue in this way, no core damage and radioactive release would have occurred.

Management of nuclear safety must depend, in part, upon reliable safety equipment. People must not interfere improperly with the performance of safety equipment.

Lesson 3: Operators and Engineers Must be Well Trained in Diagnostic Ability

The TMI operators were not adequately trained. They and the supervisory engineers did not diagnose elementary information which was adequately displayed. They misinterpreted the trends of decreasing reactor pressure and rising water level in the expansion tank. This tank is also used to establish pressure for the reactor coolant. Not recognizing that the relief valve was open at the top of the expansion tank, and thinking that there was too much water in the reactor system because of the rising level in the expansion tank, the operators stopped the flow of emergency cooling water.

During the 104 minutes that it took for the water to boil away and finally uncover the reactor core, there was sufficient time for operators and engineers to diagnose the instrument readings and learn that stopping the cooling water was slowly leading to a reactor accident with significant radioactive release. However, the operators had not been trained to recognize these conditions, even though such conditions had been predicted years before and had occurred at other reactors. Neither the operators nor the engineers recognized that the reactor coolant was boiling away. They were observing the reactor coolant pressure and temperature, but they did not understand the significance of this information. If they had simply injected enough emergency coolant to keep the water temperature below the boiling temperature, which is determined directly from the pressure, the accident would not have occurred.

Not much emergency cooling water was needed at that time. Although the emergency coolant was injected automatically at a flow rate of 3,800 liters/min before the operators stopped the flow, a coolant flow as low as 530 liters/min would have prevented core uncovery.

For adequate management of nuclear safety, operators must be properly trained. But training to respond to what we think might happen is not sufficient. There must be far better education, training, and understanding, and there must be more careful selection of qualified people. Operators and engineers must have sufficient diagnostic ability to deduce and respond to events that may not be the same as those prescribed in the training programs.

This lack of adequate diagnostic ability was also exhibited at TMI by the regulatory agency, as is discussed under Lesson 7. A new approach to training and qualification of all people in the management of safety is required. At each level some degree of diagnostic ability is necessary and essential. The management of nuclear safety demands that this ability exist and that it be periodically and continually updated and verified. This is the most important lesson to be learned from the TMI accident.

Lesson 4: There Must be Adequate Procedures

The known leaks in the TMI relief valve caused the operators to ignore the procedures and symptoms which should have told them that the valve was open. Some leaks in normal operation could be tolerated if the procedures for an accident were properly written to take this into account.

There were written procedures at TMI to aid operators and engineers to respond to a loss of coolant accident, including a small-break accident such as a stuck-open relief valve. However, the procedures for the small-break accident were not adequate and could not be used. They mistakenly required symptoms which did not exist during the TMI accident, including loss of off-site power and loss of an emergency generator.

The TMI operators were misled by a legal specification which emphasized that the pressurizer expansion tank should not be allowed to fill with water. They did not understand that this written requirement did not apply under transient and accident conditions.

These inadequate and incorrect TMI procedures contributed to the mistakes made by the operators and engineers. These procedures had been reviewed and approved by plant management and by the regulatory agency, but the deficiences had not been recognized.

Management of nuclear safety requires careful attention to procedures, including their adequacy and their comprehension by the operators and engineers.

Lesson 5: The TMI Safety Equipment Performed Better than had been Expected

At 104 minutes after the operators stopped the flow of emergency cooling water at TMI, the water had boiled away enough to uncover the reactor core. Water continued to boil away and during the next 98 minutes, until the operators restored the flow of emergency cooling water, all but the lower third of the fuel was uncovered. The uncovered fuel became so hot from fission product heating that the zirconium cladding reacted with water vapor, which caused the fuel to heat even more. About half of the zirconium was fully oxidized. Enormous quantities of hydrogen gas were formed, most of which escaped into the containment and burned. The oxidized cladding and the oxide fuel crumbled into a rubble of cracked material, resting on top of the still intact fuel at the bottom of the core. None of the uranium dioxide fuel is calculated to have reached its melting temperature of 2,760° C, although it is likely that some of the fuel dissolved locally in the hot zirconium.

Millions of curies of radioactive noble gases, iodine, cesium, strontium, and other fission products were released from the fuel. The estimated[2] curies quantities of these fission products originally in the fuel and the fractions released from the fuel, to the building air, and to the atmosphere are shown in Table 1.

Table 1. Radionuclide inventories and releases at Three Mile Island.

Radionuclide	Core inventory, curies	Fraction of core inventory released		
		from fuel	to building air	to atmosphere
^{85}Kr	9.6×10^4	0.6	0.6	~ 0.1
^{133}Xe	1.4×10^8	0.6	0.6	~ 0.1
^{131}I	6.5×10^7	0.5	7×10^{-5}	2×10^{-7}
^{137}Cs	8.4×10^5	0.5		0
^{90}Sr	7.7×10^5	$< 7 \times 10^{-4}$		0
^{140}Ba	1.4×10^8	2×10^{-3}		0
Assumed by NRC for reactor licensing[6]				
^{85}Kr + ^{133}Xe		1.0	1.0	
^{131}I		0.5	0.25	

During the accident most of the radioactive noble gases were released from the fuel. The ^{133}Xe reaching the atmosphere was the main source of the small radiation dose to the public. The radionuclide usually considered the most hazardous in accident analysis is radioiodine. It is significant that most of the radioiodine released from the fuel remained in the coolant water, and only a relatively small fraction, less than 0.01%, escaped as a gas into the containment and auxiliary building.[1,3] The amount of iodine that escaped from the plant was between 10 and 20 curies, small enough to make only a negligible contribution to the radiation dose.

Earlier assumptions and studies by the former U.S. Atomic Energy Commission[4] and by the Nuclear Regulatory Commission[5] have suggested and assumed far greater core damage and greater releases of radioactivity from the fuel and into the containment air when a core is overheated under such degraded cooling conditions. The NRC study assumed that core melting and loss of containment would occur in a small-break accident sequence involving operator error, similar to the initiating sequence at TMI. It predicted far greater releases from the fuel and to the atmosphere than occurred at TMI.

For purposes of reactor licensing, NRC assumes[6] that an accident with highly degraded cooling conditions would release half of the inventory of radioiodine from the fuel and half of that to the containment air. The fraction released to building air at TMI was about 10,000 times less.

Equipment failures at TMI were not the real cause of the accident. Paradoxically, the accident was a demonstration that the safety equipment was effective and that it could perform better than expected. It demonstrated the success of the multiple-barrier defense-in-depth approach towards reactor safety. The accident also indicated many areas wherein equipment modifications can result in further improvements in safety of existing and future reactors.

Lesson 6: The Operating Company and Industry should be Prepared to Manage Recovery from an Accident

The operating company was seriously unprepared for an accident at TMI, as can be seen from the above examples. The company was not prepared to respond to the needs for informing the government and public. The company had made no preparations for recovering from an accident, but it quickly assembled an industrial team that had excellent knowledge and diagnostic ability. The industry was not prepared to handle the cleanup of radioactivity at Three Mile Island. Although basic technology was available at government laboratories, new process design and equipment was required to implement the cleanup from the reactor accident.

Management of nuclear safety must include better preparation for the recovery and cleanup from an accident.

Lesson 7: A Reactor Accident can be Serious even if Radiation Health Effects are Negligible

The U.S. Nuclear Regulatory Commission (NRC) provided vital and needed communication with the public and with local government following the accident. However, serious technical mistakes made by NRC caused an accident that caused only negligible health effects from radiation to escalate into an incident with serious consequences. Two days after the accident NRC misinterpreted information concerning a transfer of radioactive gas back to the containment. It made an incorrect calculation of new radioactive release, made an incorrect conclusion from plant surveillance data, and recommended evacuation. The governor decided not to accept the NRC advice, but public fears were escalated.

Two days after the accident NRC learned from the operating company that a large volume of hydrogen at 60 atm. partial pressure remained in the reactor vessel and coolant system. An NRC Commissioner reasonably asked about the possibility that oxygen could be generated by water radiolysis and could accumulate enough to cause a hydrogen explosion. The NRC staff consulted did not know that in normal operation of a pressurized-water reactor hydrogen is added to the reactor coolant at a partial pressure of about 0.5 atm. so that no oxygen will be formed by radiolysis, even under the intense radiation field of normal operation. The NRC staff was correctly informed both by the operating company and by the reactor designer that the hydrogen pressure of 60 atm. was more than sufficient to supress oxygen formation, and that an explosive mixture could not form.

However, the NRC staff made their own calculations which ignored the effect of dissolved hydrogen to prevent radiolysis. They accepted similar incorrect calculations from others, ignored data in their own files concerning oxygen generation in operating reactors, and mistakenly concluded that the hydrogen in the reactor vessel would soon become combustible. Three days after the accident NRC announced this mistaken information to the public, along with concerns about possible rupture of the reactor vessel and containment and possible melting of the reactor core. Public fears were highly escalated, and there was no ready source of information to correct the errors made by NRC. Shortly thereafter NRC learned again from industrial groups that oxygen could not form and that the hydrogen bubble could not explode. They informed the public of this correct information a few days later.

The President's Commission and its investigative staff concluded that the public fears, hysteria, and trauma were a serious health effect from the accident, though short-lived. TMI tells us that a reactor accident which causes only negligible health effects from radiation can become serious because of fright and hysteria.

The management of nuclear safety requires that the regulatory agency have sufficient technical knowledge and diagnostic ability in the system that it is regulating. The regulatory staff should be able to diagnose events important to public safety and should be able to obtain and interpret diagnostic information from reliable sources when needed. The regulatory agency should recognize the need for this ability within its organization and should develop and maintain this necessary competence. The role of the agency in the event of an accident needs to be defined.

Lesson 8 : Consequences of Possible Failures Should be better Understood

If the TMI operators had not finally restored the cooling water, after the core had been uncovered for about one and a half hours, fuel melting could have begun in another hour or so. The reactor was that close to melting during the day of the accident. The possibility of even later melting was raised by NRC in connection with its hydrogen-bubble mistake three days later. There was no reliable information to guide the operating company, the regulatory agency, and the public as to what to expect in the event of melting. Lack of such information encouraged much incorrect speculation and public fear.

Studies of the consequences of a melted core were terminated years ago. Even though the Rasmussen study had concluded that melting could occur under some accident conditions, it was evidently concluded by AEC (Atomic Energy Commission) and NRC that designing to protect against the large-pipe-break accident would preclude any chance of melting. However, TMI tells us that small malfunctions and human errors can compound into an accident perilously close to fuel melting.

Recent analyses carried out by the industry and by the staff of the President's Commission show that if the flow of cooling water had not been restored at TMI a molten mass of uranium dioxide fuel at 2,760°C would have slowly melted through the bottom of the reactor vessel and spilled onto the containment floor. This would have initiated the automatic flow of water from the other emergency cooling systems for the reactor and in the containment. The deluge of water contacting the molten fuel would have caused steam explosions, but they would not have damaged the containment. The mass of fuel on the containment floor would have been adequately cooled. It was concluded that even if this mass of fuel was not water cooled, the heat-dissipation capacities of the 4-meter-thick concrete floor and the base pad would retain the core for several days. It was concluded that the core would eventually refreeze, even if not permanently water-cooled, and that the containment would have survived to prevent dangerous radioactive release to the public.

The management of nuclear safety requires thorough, factual analysis of events such as this, to provide a better understanding of consequences and for action planning in the event of emergencies. More and better factual information can displace much of the fear and speculation that interfere with the management of nuclear safety.

Lesson 9: Accidents Must be Expected and Planned For

The TMI experience tells us that nuclear safety cannot be adequately managed on the basis of assumptions and beliefs that reactor accidents cannot and will not occur. Indeed, TMI confirmed the calculations in the 1975 study by Rasmussen,[5] who predicted that a small-break accident with operator error would occur by about the time of TMI, within the error bounds of the prediction.[2]

A lesson from TMI is that a safe low-risk system still has possibilities for accidents. Every technology imposes a degree of risk upon society, both in its routine operation and in the occurrence of accidents. Over a long enough time period even low-probability accidents may occur. We must be prepared for these accidents. The only approach that can result in public safety is to manage nuclear safety so that accidents have sufficiently low probabilities and low consequences.

Lesson 10: We Must Learn from Experience

One of the most disturbing lessons from the investigation of the TMI accident is that industry and government have not given sufficient attention to earlier experiences that were direct warnings of the possibility of the TMI accident. Events similar to those that occurred during the first few minutes of the TMI accident, when the operators were becoming confused by the rising water level and falling pressure, had happened twice before. In 1974 a pressurized-water reactor in Beznau, Switzerland, constructed by a different reactor supplier, experienced a similar initial transient event, but the operators diagnosed the situation early and terminated the event before damage occurred. This event was known to parts of the U.S. industry and government but was not made known to U.S. operating companies like TMI.

In 1977 a reactor at the Davis-Bessee plant in Ohio, almost identical to the TMI reactor, also experienced a turbine trip, a stuck-open relief valve, decreasing reactor pressure, and rising level of water in the expansion tank. There, as at TMI, the operators mistakenly stopped the flow of emergency cooling water. Fortunately, after a short period the cooling water was restored. The event was investigated and reported by representatives of the NRC and the reactor supplier, but neither organization acted directly upon this information to avoid a similar happening at other places. These are alarming examples of inadequate management of nuclear safety.

Further examples are equally alarming. In 1977 an engineer at a U.S. utility company predicted that an open valve in the expansion tank could cause the rising water level and a falling pressure, and he was concerned that this could confuse operators. His predictions were made known to the supplier of the TMI reactor, to the U.S. Advisory Committee on Reactor Safeguards (ACRS), and to the NRC. Although the ACRS did send, through NRC, an inquiry to a reactor owner concerning this possibility, neither the ACRS nor the NRC followed up after no response was received.

In 1971 a different U.S. reactor supplier predicted the possibility of a confusing rise in water level in the expansion tank when a relief valve remains open. It notified some reactor operators and the NRC. The NRC did not notify other reactor suppliers and operators of this problem.

If any one of these several earlier events had been properly diagnosed and understood by the other operating companies as a precursor of a likely accident, the TMI accident would probably not have occurred.

The management of nuclear safety should include careful and thorough diagnosis of operating events with the view that they may be precursors of new kinds of accidents that have not previously been analyzed or encountered. There must be a system to promote the dissemination of this information to others who can aid in the diagnosis and who may be affected by the results. The past system in the U.S. is woefully inadequate. It can be corrected only if both industry and government accept responsibility that the understanding and dissemination of such information is vital to maintaining safety. This requires competent and qualified people working towards this goal, and it requires the full support of management and government. Better dissemination of information will require greater penetration of the proprietary barriers of industry and the adversary climate of government.

Lesson 11 : Emergency Planning should be Realistic

The TMI emergency plans requiring public action and evacuation were based upon unrealistically high radiation exposures specified by the regulatory agency for the large-pipe-break accident. There was little preparation within the community and state for responding to such emergency. They were poorly prepared to diagnose the public-health threat from the radioactive releases. From the TMI experience we learn that there must be contingency plans based upon different kinds of accidents, with different time scales and different radiation exposures.

Management of nuclear safety requires more comprehensive and realistic planning for emergency response, it requires centralized authority for authorizing and managing the off-site emergency response, and it requires an adequate communication system.

Lesson 12: There should be Adequate Preparation for Informing Government and the Public

The TMI experience demonstrates that responding to the needs of the government and public for information during and after the accident can be a most difficult problem for the operating company. The government role of informing the public is equally difficult. The person to provide this information must speak with knowledge and authority. Past approaches to providing public information are not adequate.

News media must be competent to understand and report the information. They must be able to distinguish fact and informed opinion from speculation. TMI shows that careless and incorrect information, particularly when originating within the authority of industry or government, can exacerbate an accident of negligible radiation hazard to one of serious public concern.

Preparing for proper and adequate information transfer is a major problem in the management of nuclear safety. It may become one of the most difficult responsibilities of nuclear plant management and of the government.

Lesson 13: Detailed Regulations do not Assure Nuclear Safety

The TMI investigation revealed excessive reliance by the regulatory agency on an excessive number of detailed written regulations which seem to specify how a reactor should be designed and operated for safety. Writing these regulations and evaluating nuclear plants against these regulations has been the preoccupation of the regulatory agency. Developing the massive written responses to these regulations for each plant has occupied much of the best talent within the nuclear industry. There has resulted the complacent assumption that safety results once these regulations are complied with. The investigation by the President's Commission concluded that this regulatory approach interferes with safety.

There are many other problems with the current U.S. regulatory approach. The regulations themselves have not been developed with an adequate focus on what are the most important elements of safety and how they are to be achieved. These regulations and guidelines have been developed without quantified safety goals and objectives. Many safety concerns have been postulated and acted upon without adequate evaluation of their importance to safety or of their consequences. Present and past approaches to U.S. regulation do not exhibit a reasonable determination of priorities or allocation of resources in proportion to the estimated risk to the public. A disproportionate effort on the part of government and industry has been required for some issues which have only a marginal impact upon risk. Many of the requirements are arbitrary and are mandated without valid technical back-up or value-impact analysis.

Many of these problems reflect lack of determination by the regulatory agency and by government of the real purpose and goals of nuclear-safety regulation. It also reflects a staff and management with too little practical experience in designing and operating equipment similar to that which they regulate.

Another problem in the regulation of nuclear safety is the stifling adversary approach. The existing U.S. process inhibits the interchange of technical information between the NRC and industry. Lack of such interchange was one of the primary contributors to the TMI accident.

In addition to insufficient evaluation of operating experience, as has already been discussed, the present regulatory system shows a lack of comprehensive approach to the entire plant. A large proportion of the regulatory staff are specialists focusing upon narrow topics. There are relatively few systems engineers with the experience and qualifications to integrate individual safety features into an overall concept and to place issues in perspective.

Many of the problems of diagnosing events relevant to reactor safety can be attributed to the overwhelming emphasis within the regulatory agency on the use of conservative models and assumptions. It is important to ensure that expected operating conditions are sufficiently removed from failure limits. However, this can be achieved by means other than intentionally distorting the equations of physical phenomena, which must remain tools for interpreting and understanding what is really happening. Safety is compromised when responsible professionals rely instead upon equations, models, and computer codes with buried and forgotten distortions, made non-realistic for the purpose of incorporating design margins.

Management of nuclear safety requires a competent regulatory approach which is aimed at identifying the main goals and problems of safety and which is implemented by realistic analysis and diagnosis.

Lesson 14: The Operating Company is Primarily Responsible for Nuclear Safety

Even the most competent of regulatory agencies cannot assure nuclear safety. Primary responsibility rests with the operating company, together with the engineering firms and equipment suppliers that design and build the plant. The operating company must expend sufficient manpower and energy to go beyond fulfilling the requirements of the regulatory agency. It and its suppliers must have qualified manpower to attend to all problems of nuclear safety, whether or not they are required by regulations. They must take the initiative in innovation of new ideas and procedures to maintain and enhance nuclear safety.

3. Concluding Remarks

These are some of the principal lessons learned from the Three Mile Island accident that are important to the management of nuclear safety. These remarks are critical of many parts of the U.S. nuclear reactor industry and nuclear regulatory agency. However, these criticisms should not obscure the fact that in over 500 reactor-years of commercial nuclear power operation in the United States there has still been no identifiable effect upon the physical health of the public due to radiation and radioactivity from nuclear power plants. This outstanding record has been achieved by the industry and the Nuclear Regulatory Commission, the parties that have been criticized, and under the system that has been criticized.

The extent to which government and industry in the U.S. are solving the fundamental problems revealed by the TMI investigation cannot yet be evaluated. NRC has reacted quickly by slowing down licensing, while it specifies detailed requirements to cover the immediate faults and malfunctions which occurred during the TMI accident. Nine months after the accident NRC issued a detailed draft of its proposed post-TMI action plan. Over 200 items were proposed. Industry, ACRS, and others have pointed out the difficulty of complying with all items on a short time scale and have recommended that a distinction be made between those tasks that are important to safety and those that are less important.

Some reorganization within NRC has occurred, including the formation of a new group to monitor and diagnose operating events. The President has adopted most of the recommendations of his Commission and has proposed legislation to increase the management authority of the NRC Chairman. He has created a Nuclear Safety Oversight Committee.

The electricity utility industry responded to TMI by forming its Nuclear Safety Analysis Center (NSAC) to study the accident and nuclear plant safety questions. Technical analyses by NSAC, as well as analyses by individuals in the Department of Energy laboratories, were of considerable benefit to the President's Commission in carrying out its investigations.

The electric utility industry later created the Institute of Nuclear Power Operations (INPO), aimed at solving one of the central problems revealed by the TMI accident, the problem of qualified people to operate and manage nuclear power plants. INPO is to establish standards of quality and competence, and it is to regularly evaluate operating practices at nuclear plants. It is to set criteria for training programs and instructors. It will review and analyze operating experience, emergency preparedness, and the human aspects of plant systems design. The undertaking of INPO is a massive and difficult program of self improvement by the electric utilities that operate reactors. The intensity with which the industry has initiated this venture indicates their intent that it be successful. However, it will not be easy for INPO to influence changes in the management approach to safety, where needed, among so many different member companies. Nor will it be easy for a similar operation to be instituted for the government regulatory agency.

The most important lesson to be learned from the Three Mile Island accident is that the industry and regulatory agency badly need people at all levels who are qualified for their jobs. They must understand the equipment and system that they are operating, building, managing, or regulating well enough both to perform routine tasks and to make responsible action decisions in the case of emergency.

This problem of qualified people can be solved in part by more intensive training and by enforcing more stringent requirements of education, training, and experience. However, detailed requirements of course work and hours on reactor simulators are no more a safety solution than are the detailed specifications of equipment by the regulators. Individuals who hold these positions should also have the understanding and diagnostic ability essential to implement their responsibility in nuclear safety. Continued development and updating of diagnostic ability should be emphasized.

We need to emphasize higher professional standards of excellence and individual responsibility among the engineers and scientists who staff the industry and regulatory agencies. This is where a great university like the University of Tokyo can make such an important contribution through its Department of Nuclear Engineering. Your graduates are not reactor operators, but they will be in key positions in government and industry. Implementing the high professional standards that they will have learned from you and setting similar standards for others is one of the best means of avoiding the many human failures associated with accidents like Three Mile Island.

References

1. J. G. Kemeny, G. Babbitt, P. E. Haggerty, C. Lewis, P. A. Marks, C. B. Marrett, L. McBridge, H. C. McPherson, R. W. Peterson, T. H. Pigford, T. B. Taylor, A. D. Trunk: "The Report of the President's Commission on the Accident at Three Mile Island" (1979).
2. K. H. Ardon and D. G. Cain: "TMI–2 core Heat-up Analysis," NSAC–24 (1980).
3. W. R. Stratton *et al.*: "Staff Reports to the President's Commission on the Accident at Three Mile Island: Reports of the Technical Assessment Task Force, Vol. II" (1979).
4. J. J. DiNunno, F. D. Anderson, R. E. Baker, R. L. Waterfield: "Calculations of Distance Factors for Power and Test Reactor Sites," TID–14844, U.S. AEC (1962).
5. N. Rasmussen *et al.*: "Reactor Safety Study: An Assessment of Accident Risks in U.S. Commercial Nuclear Power Plants," WASH 1400 (1975).
6. U.S. Federal Register, 10 CFR 100.

Nuclear Safety: Its Achievement in Perspective

Armin JAHNS

1. Introduction

Commercial use of nuclear power production in Germany started in 1961 with the commissioning of the 17 MWe boiling water reactor VAK. Currently, 14 reactors are in operation with approximately 9 GWe, 11 are being constructed with about 13 GWe, and 10 are projected with another 12 GWe. This would roughly meet the energy goals set forth by the Federal Government which aim for 30 GWe by 1985.

The peaceful use of nuclear energy in Germany has faced a handicap from its very beginning: the lack of a large number of suitable reactor sites, mainly due to the fact that our geography shows only a small number of the needed coastal areas or big rivers to provide for cooling water. Furthermore, a demographic analysis reveals that these relatively few suitable sites even show an increase over the already high average population density. So most of the sites that are favorable in their energy-producing aspects must be excluded due to aspects of environmental protection or safety considerations.

From these few thoughts one may see that nuclear safety was the task to focus on to render possible nuclear energy production in Germany.

2. Concepts of Nuclear Safety

In conventional technology safety concepts and measures have been based, at least to a substantial extent, on experience or what you might call "trial and error." Nuclear energy could not follow this path for obvious reasons. Here, theoretical considerations led to a safety philosophy whose effectiveness again was theoretically examined by numerous and very sophisticated incident analyses which analyzed the consequences of certain representative abnormal courses of events. Again, it was contemplated how these events might be precluded or how their consequences could be confined to an acceptable limit. These more theoretical efforts formed the basis

Geschäftsführer, Reaktor-Sicherheitskommission (RSK), FRG.

for an internationally applied safety concept called "defense in depth." Such a "multilevel" safety concept precludes to a high degree the release of great amounts of radioactivity by different, independent requirements.

First Safety Level: Operation

All components of the plant are designed, manufactured, and constructed in such a manner that malfunctions are most unlikely.

Second Safety Level: Abnormal Operation States

Should such disturbances occur nonetheless, the reactor protection system prevents them from leading to dangerous states of the plant by automatically inducing such protection measures as reduction of power output or shutdown of the reactor.

Third Safety Level: Incidents

During the lifetime of all plants the occurrence of incidents cannot be completely excluded. The safety concept therefore incorporates safety systems to interfere in a course of events to sustain the integrity of the remaining retention barriers and thus reduce the consequences of such incidents. These "engineered safeguards" have to perform with exceptional reliability, which is accomplished by the following design criteria:

Redundancy;
Diversity;
Logical linking of the impulses;
Physical separation;
Single-failure criterion.

These precautionary measures have accomplished such a high degree of safety that during 25 years of the peaceful utilization of nuclear energy, no person has been harmed by its inherent danger. In the Federal Republic of Germany, since 1975, 789 safety-relevant occurrences were reported from the utilities of the 16 nuclear power plants existing in that time period; 74 of these note the release of radioactivity out of the normal operation state into the environment, of which 24 cases exceeded the authorized release limits. It must be noted, however, that in none of the occurrences and incidents were the annual release limits exceeded, and dose assessment for the environment showed values far below the dose limits given by the Radiation Protection Ordinance. Even the incidents at Gundremmingen (January 31, 1977: overfeeding of the primary coolant system) and at Brunsbüttel (June 18, 1978: release of fission product in the machine house and through the stack) show only about 1 % of the authorized release limits.

All this proves that the prevailing safety philosophy and safety concepts fully serve their purpose to reduce the risk associated with the use of nuclear energy to an acceptable value.

3. Licensing Procedure

A short outline of the licensing procedure in Germany is as follows. Under the Atomic Energy Act, the states, on behalf of the Federal Government, are responsible for licensing nuclear facilities (such as nuclear power plants and reprocessing plants) and, in addition, supervise the construction and operation of such plants. The licensing authorities in each case are the supreme state authorities. Their activities are subject to supervision by the BMI (Bundesministerium des Innern) with respect to legality and expediency. While the state ministries employ experts for the technical assessment of plants and for quality checks, the BMI is advised by the RSK (Reaktor Sicherheits-kommission), among other bodies. In this task the commission assesses the safety concepts of nuclear power plants and fuel cycle facilities and in this way contributes significantly to the definition and advancement of the state of the art. In its advisory capacity the RSK concentrates mainly upon novel problems and questions of fundamental importance. The RSK judges whether the technical preconditions of licenses of nuclear facilities are met in accordance with the requirements under the Atomic Energy Act. As a result of its deliberations RSK issues recommendations which constitute an important basis for the directions ordered by BMI to the supreme state authorities.

4. New Guidelines

The RSK has a profound influence on reactor safety technology and, through its guidelines and specific recommendations, contributes to the high safety standard generally recognized. In early anticipation of the number of nuclear power plants that would be built, the RSK began to compile general safety criteria in so-called guidelines. In the review of the earlier PWR-guideline version of 1974 RSK identified several areas which required greater attention in order to incorporate the progress gained in reactor safety. The more important sections in the new guidelines (published Jan. 1979) were requirements on the following four categories.

4.1 Basic Safety
"Basic safety" of all pressurized components and systems is necessary for the mitigation of incidents. The "General Specification of Basic Safety" lays down principles on:
High-quality material characteristics, in particular toughness;

Conservative restriction of stress;

Prevention of stress peaks through optimal design;

Assurance of the application of optimized manufacturing and testing technologies;

Detection and assessment of possible faulty conditions.

If the requirements contained in the present General Specification are complied with, the components will attain a basic safety level that will preclude any disastrous failure of a plant component resulting from faults caused by the manufacturer.

4.2   Control of Small Leaks

To cope with small leaks the following assumptions shall be made or design conditions complied with:

1. Components and systems required in addition if small leaks occur (*e.g.*, emergency feedwater pumps, secondary blowdown station as well as their activation circuits) shall be considered as sub-systems of the emergency core cooling and residual heat removal system, for which the design requirements are contained in the present chapter.
2. The failure of station service power supply shall be postulated for the incident analysis.
3. If a direct pressure reduction in the secondary system is required to cope with the incident, it should be done automatically.
4. The available water supply for emergency injection and for the borated water storage tanks shall be sufficiently conservative.

4.3   Hydrogen Limitation in Containment

To prevent any explosion or fire hazard inside the containment, efficient countermeasures are required to control the formation and release of hydrogen after a loss-of-coolant accident:

1. A monitoring system must be installed to render reliable information about the local and temporal distribution of hydrogen within the critical areas in the containment even under incident conditions. The attached analyzing system must be capable of measuring the activity contained within the collected gas samples. The measuring points must also allow for a measurement of the temperature distribution which might inhibit the homogeneous mixing of the containment atmosphere.
2. If it cannot be proved, by adequate calculation methods, that locally confined over-concentrations may occur, active features must be installed to assure sufficient forced flow mixing.
3. Likewise, active measures must be provided for (such as recombiner

connections) if an integral volume concentration of above 4% cannot be excluded.

4.4   Electrical Equipment in the Safety System

In order to assure the functionability of the reactor protection system, criteria were formulated not only for the system itself but for all electrical equipment of the safety system.

1. All active and passive electrotechnical components of safety-relevant devices must retain their reliable functionability under all plant operation modes.
2. The safety-relevant parts of the station power service and of the power grid connections have to meet the same availability standards as the emergency power supply.
3. An incident course instrumentation shall be designed in such a way that the variables of state selected for the determination of an incident course will be documented in a clearly arranged form and in the proper chronological order.
4. An incident sequence instrumentation shall be designed in such a way that the data which, after the occurrence of any event that may lead to an increased release of fission products to the environment of the nuclear power plant, are of decisive importance to safety with regard to the assessment of plant safety, the efficiency of the safety system, and the decision on emergency measures, will be indicated and documented reliably and with a sufficient degree of accuracy.

5.   Impacts of the Accident at TMI-2 Plant

From these examples it can be seen that the main problems which occurred during the accident at the Three Mile Island (TMI)-2 plant had been identified and regarded in these safety requirements before the accident. Immediately after the accident, RSK started an investigation to determine if such a sequence of events or a similar one could be possible in German PWRs. It was found that the German reactor design differs from the American in some relevant points. These differences would correct human error more easily and block disturbances and incidents earlier:

a. Because the emergency feedwater system of German PWRs is more redundant and the redundant trains are non-intermeshed, the connection to the four steam generators is not interrupted simultaneously in case of maintenance or repair.
b. Due to a larger secondary coolant volume the steam generators would not have evaporated under similar conditions.
c. The pressurizer relief valve in German plants is secured by an initial

valve so that the primary circuit would have been isolated and saturated conditions would not have been reached.

d. The lifting pressure of the emergency high-pressure injection system lies below the response pressure of the pressurizer relief valve so that there may be no need to turn off the automatic emergency cooling system.

Apart from this more comparative investigation, RSK analyzed the safety systems of those reactors which are currently in operation or under construction to determine whether their safety design had to be back-fitted to cover TMI-like incidents. No short-term measures had to be taken, while the necessity for mid- or long-term measures required closer investigation.

The main factors which determined the course of the accident were the following:

A falsely open relief valve which acted like a small leak in the primary circuit;

Failure of the staff to recognize the critical operating conditions.

It can be seen from these factors that possible findings would touch on improved data processing and operating instructions to better detect the state of the plant, further intensified education of the staff, and in some points the field of engineered safeguards.

Following were the measures recommended:

a) Short-term:

Intensified education of the staff with particular emphasis on the control of abnormal occurrences between normal operation and incident;

Reliable actuation signals for shut-off valves.

b) Mid-term:

Better information on the state (boiling state) of the primary coolant;

Installation of an accident-resistant hydrogen monitoring system;

Automatic containment isolation by activity signals from monitoring devices within the containment;

Connections for hydrogen recombiners;

Improvement of air recirculation and ventilation filter systems;

Improved reliability of the spent fuel storage pool cooling system; in German PWRs the pool of spent fuel elements is inside the containment and therefore not accessible in case of a LOCA.

c) Long-term:

Assessment of necessity of a fan system to avoid build-up of critical hydrogen concentrations.

In summary, it can be concluded that the evaluation of the TMI accident did not reveal the necessity for any drastic changes in the safety design of German nuclear power plants. So on the more technical side TMI was not the milestone in safety philosophy that it could have been.

The situation proved a little different from the political and public viewpoints. The authorities required a re-review of the safety philosophy and its underlying prerequisites and assumptions.

6. German Risk Study

One requirement for the issuance of a construction license is the proof of adequate precaution against selected postulated incidents whose radiological consequences may not result in individual exposures exceeding a set limit of 5 rem whole body dose (15 rem thyroid dose). These events are listed in a list of representative "design basis incidents" whose occurrence must be postulated during the lifetime of a plant with a probability of 10^{-4} to 10^{-2} per year. One may, of course, conceive of events which lead to exposures exceeding the allowable limit. The occurrence and sequence of such events, however, are either not foreseeable or in practical judgment so unlikely that further measures against such hypothetical events are considered unnecessary. The German risk study (Deutsche Risiko Studie: DRS), comparable to the American WASH-1400 study, assessed the risk resulting from such events. In short, the findings did not differ significantly from the American analysis. Thus both studies state that the containment vessel of a nuclear reactor would greatly reduce with a high probability the consequences of a core-melt accident.

The potential hazards deriving from the "accidents," as they are termed in German, are accounted to the so-called residual risk associated with the peaceful use of nuclear energy as well as with other manmade technologies. It is commonly accepted that any precaution against design-basis incidents simultaneously reduces the probability of accidents since these are the results of uncontrolled design-basis incidents. Any improvement in the safety system therefore diminishes the residual risk. The continuous incorporation of an evolving new state of the art in nuclear safety into the safety requirements is considered the most effective countermeasure against such accidents. On the other hand, one may, of course, contemplate devices which are apt to mitigate the consequences of such an accident and prevent a massive release of fission products to the environment, no matter how unlikely such a release may be.

The work on the DRS not only allowed for an assessment of the potential risk of a nuclear power plant but it also identified weak points in the reference system in the course of the analysis. In close cooperation with the manufacturer and operator of the reference plant a number of measures were detected that could improve the availability and reliability of the safety system and, thereby, reduce the probability of a severe core-melt accident.

As a result of the recent discussion, measures for incident prevention and consequence limitation that have been recommended. These are as follows.

Some of these measures will be introduced. Since the possibility cannot be excluded presently that some measures may prove safety-related advantages as well as disadvantages, RSK will decide upon these measures after the initiated investigations have been concluded.

Measures for consequence limitation also have been and are being subject to extensive study in the Light-Water-Reactor Safety Research Program conducted in Germany, ever since the utilization of nuclear power production started.

A.   Measures for Incident Prevention

Measurement of pressure and temperature of the primary system to generate a signal to indicate whether the primary system is in a sub-cooled or saturated state;

Remotely controlled degassification of the reactor pressure vessel or of the primary system;

Improved temperature monitoring of the reactor core;

Provision of instrumentation to indicate the water level in the pressure vessel;

Increase of the admissible pressure in the secondary coolant circuit for better control of breaks of steam generator tubes (requiring stronger pipes up to the turbine bypass station);

High-pressure injection from the sump;

Further development of programs for simulator training (different development for disturbances and accidents);

Improved display of the state of the primary system in the control room (electronic data processing);

Fast access to the operation manual by the use of electronic data processing;

Use of computers for accident and accident consequence analysis;

Investigation of the natural convection within the primary system in case of two-phase conditions (including the dynamics of the pressurizer and the presence of non-condensible gas).

B.   Measures for Consequence Limitation

Heat removal from the containment by an outside spray system. This would prolong the time for emergency response activities. Its technical feasibility, however, has to be proved;

Venting of the containment through filters or into a pressure-suppression system. Further investigations and an evaluation of the merits are required;

Development of a dry reactor building basement to act against the melt-through of the core and to diminish the production of gas and steam. A necessary relocation of the containment sump should be investigated;

Underground siting. Further research is necessary.

It should be noted that the RSK has recently assessed underground siting of nuclear power plants. It concludes that, on the basis of the existing studies, such underground plants show no safety advantage over above-ground reactors, if these engage a safety system improved by the recommendations indicated above.

In summary, it may be expected that after investigation of all of the proposed measures for incident prevention and consequence limitation there will be safety engineering improvement. Due to the already achieved high safety level, any such improvements, however, probably will be marginal. Nevertheless, RSK strongly supports this work and will work for the realization of noticeable improvements.

7. The Back End of the Fuel Cycle

There is one field of nuclear safety which is often neglected in safety considerations, to which the last part of the paper will be devoted: the management of the spent nuclear fuel.

The German Government has fully recognized the importance of the "other end of nuclear power production" and has therefore established a link between the issuance of any further licenses for nuclear power plants and proof of safe management of spent fuel (*Entsorgung*). Under the "Principles relating to the provisions to be made for the Entsorgung of nuclear power plants," the achievement of the following in the realization of the integrated *Entsorgungs*-concept shall be necessary for the issuance of first partial construction licenses:

Preselection of one or more fundamentally suitable sites, either for an external intermediate storage facility insofar as no intermediate storage is assured at the site of the nuclear power plant, or for a fuel reprocessing plant;

Positive assessment by the respective advisory bodies of the fundamental safety-related realizability of the intermediate storage of spent fuel elements at external intermediate storage facilities over a period of at least 20 years;

Continuation of the current land use planning procedure as well as progress in the exploration and development of an ultimate storage facility.

After January 1, 1985, an additional condition for the grant of a first partial construction license is that in the course of the construction of one or more fuel reprocessing plants or one or more plants for the treatment of spent fuel elements for ultimate storage without reprocessing, the preselection of a site must have been made for one of these facilities.

This new regulation was stipulated after the plans for a 1,400-tons-per-year integrated nuclear fuel cycle center had to be deferred, due to interior political obstacles. The necessity for a reasonable *Entsorgungs*-concept can easily be seen by a look at the accumulation of irradiated fuel elements. The Federal Republic of Germany currently operates 14 commercial light water reactor power stations with an installed electrical output of 9 GWe. It can reasonably be assumed that this generating capacity will increase to almost 30 GWe by 1990 and to over 50 GWe by the year 2000. This in turn means that the annual production of irradiated fuel elements will increase from the current level of 230 tons of uranium to about 750 tons by 1990. Over the same period, the quantity of spent fuel to be stored will have accumulated to over 6,000 tons of uranium.

The capacity for on-site storage will be increased at the beginning of the 1980s by the installation of so-called compact storage frames in the spent fuel storage pools. Thus the capacity of these pools will be increased considerably over the present level of about two discharge batches. Safety problems associated with the increase in activity levels or the production of decay heat are neglectable, since these are determined primarily by the last discharge.

The current nuclear waste management strategy is a manyfold concept. The demonstration plant (WAK) is of particular importance in the planning of a large-scale German plant. Since 1971, it has been used successfully to reprocess fuel elements from various experimental and production reactors with burn-up of up to 39,000 MWd/tU.

A large proportion of the irradiated fuel elements produced in the Federal Republic of Germany is reprocessed or put into interim storage in the French reprocessing plant at Cap la Hague. In addition, the contracts with France also provide for conditioning of the resultant highly radioactive waste by vitrification and its subsequent return to the Federal Republic of Germany. Irrespective of the decision on a site for the planned National Nuclear Fuel Cycle Center, it is planned to use future final storage facilities to deposit this waste. The same applies to radioactive waste from the Karlsruhe reprocessing plant. In the medium and long term the National Nuclear Fuel Cycle Center alone should be able to cope with all the fuel elements from German nuclear power stations. The planned plant with an annual throughput of 1,400 tons of uranium will encompass all areas of waste management, from fuel element storage, reprocessing, uranium- and plutonium-processing and conditioning to final storage of the radioactive waste materials in a salt dome at the site.

An alternative approach to this integrated concept plans for two regional interim storage sites, in North Rhine-Westphalia and Lower Saxony, each with a capacity of 1,500 tons of uranium. Moreover, the State of Hesse declared that it was prepared to provide a site for a reprocessing plant with an annual capacity of 350 tons.

At the same time, the direct final storage of spent fuel elements without reprocessing is being examined from the point of view of feasibility and safety. However, since there exist only the preliminary results of this concept, a final decision cannot be expected before the mid-1980s.

From the point of view of safety the current waste management strategy, i.e., the interim storage of irradiated fuel elements in compact stores for subsequent reprocessing in foreign plants, seems adequate, at least in the short term. However, in the longer term, a strategy at the national level is essential to reduce dependence on other countries in the energy sector. Without doubt the concept of the integrated Nuclear Fuel Cycle Center represents the best means of realizing this aim from a safety point of view. The plan to dislocate reprocessing and waste disposal, which has emerged for political reasons, must be regarded as less favorable because of the hazardous additional transportation of nuclear waste.

Direct final storage, particularly in recoverable form, seems to be the least attractive alternative.

8. Conclusion

The analysis of the accident at TMI has led to no remarkable change in nuclear safety concepts. With existing and technically available knowledge on nuclear safety there are no obstacles on the safety side of nuclear energy that would harm a further and intensified engagement in this source of energy. The experience with the operation of nuclear power plants in the Federal Republic of Germany has shown that the methods of reactor safety have proved to be satisfactory.

Safety-Oriented Research on Power Reactor Fuels

Yoshitsugu MISHIMA

1. What is Fuel Safety?

Nuclear reactor safety is based on the perfection of retaining fission products produced in fuel. Fuel safety study is, therefore, said to be the heart of reactor safety study.

From this point of view, the current philosophy of fuel safety is to make efforts to prevent the release of fission products formed inside the cladding into the coolant as completely as possible. Fuel assembly is so designed and fabricated as to retain fission products perfectly within cladding during normal operating conditions. In nuclear reactor safety design practice, cladding is required to be intact even in the case of transience, but the loss of containment of fission products is taken into consideration in case of an accident and fuel specialists are required to offer information on the mode and degree of failure of fuel under such accidental conditions as exactly as possible so that it can be used satisfactorily by the designer of a back-up safety system.

Emphasis in power reactor fuel design is placed, therefore, upon the assurance of containment of fission products within cladding under normal operating conditions including transience. But no attempt is made to improve fuel behavior during major accidents by sacrificing its behavior under normal and transient conditions. Abnormal conditions are those between normal and accidental conditions, and various degrees of cladding failure are expected to occur, ranging from that in which cladding is still intact to considerable loss of integrity. Fuel safety specialists are required to supply as precisely as possible information on the mode of fission product release for each failure severity. A considerable amount of information has been obtained lately on the fuel failure modes in cases of a reactivity-initiated accident (RIA) or power- cooling mismatch (PCM).

Department of Nuclear Engineering, University of Tokyo, Japan.

2. Fuel Failure Modes and Their Remedy

Based on the current philosophy of fuel design and fabrication fuel failure should not be allowed under normal and transient conditions. But as tens of thousands of fuel rods are loaded into a single larger power reactor core, it is inevitable that a few leaker fuels creep in, against which a system is added to remove fission products in the primary water. Various operating modes are possible within the range of normal operating conditions, according to the requirements of power reactor management and different modes are employed in various countries. When considered historically also, present service conditions of the LWR fuel have changed considerably from those at the time of its initial design and the numbers and modes of failure of leaker fuels have been different throughout the period.

Fuel rods containing inherent defects due to insufficient quality control actually existed about 10 years ago. In the case of rods fabricated based on improper specifications due to insufficiency of knowledge in such areas as moisture control, hydride failure occured during service. But such errors have been eliminated due to technological improvement. Recently, as the results of efforts to enhance power density based on the reduction of margin below thermal limitations, the thermal limit may be exceeded in a transient and this gives rise to the occurrence of leaker fuel. When a major cause of the failure have been identified to be pellet-clad-interaction (PCI), remarkable improvement of fuel performance has been made through the reduction of linear heat rating by decreasing fuel rod diameter and by adoption of the voluntary limitation in power-up rate as PCIOMR on the part of the reactor operator. The amount of leaker fuel has decreased remarkably especially in Japan. The fraction of leaker fuel throughout the world has come down to 4 or 5 rods per 10 thousand, whereas it has been 1 or 2 per 100 thousand in Japan since 1975: one order of magnitude lower.

3. Deformation of Fuel Rods in Service

Collapse, which occurred about 10 years ago in PWR fuel rods, has been eliminated due to the adoption of a procedure for fabricating less densifiable pellets as well as the adoption of internal prepressurization.

Bowing is not necessarily connected to fuel failure, but as it may increase failure probability, limitation of further use of bowed fuel has been imposed by the Japanese Regulatory and the removal of bowed fuel has been recommended when more than 85% gap closure is anticipated during the next cycle of operation. This has been known as the "85% criteria."

The cause of the bowing is judged by myself to be attributed to the excessive strength of the supporting grid springs in the Westinghouse-type PWR fuel assembly, rather than to the uneven thickness of cladding tubes or others. I therefore recommended its reduction. After this reduction was made, considerable reduction has been experienced in the number of significantly bowed fuel rods of more than 50 % gap closure, and no PWR fuel assembly has been removed after 1977 due to violating the 85 % critieria.

Lately, a temperature increase of about 60° C at the contact position has been verified on contacted PWR fuel rods due to bowing through a cooperative experiment among Japanese electrical utilities and reactor manufacturers, using a newly designed forced circulation rig of IFA 510 in a Halden reactor.

4. Fuel Behavior in Accidents and Japanese ECCS Criteria

Fuel behavior during a major accident was emphasized in the early days of fuel safety study, since the largest amount of fuel would fail to the greatest extent in such an accident. Thus experimental study has been carried out everywhere on the burst behavior of a cladding tube after LOCA as well as on the oxidation behavior of zircaloy cladding tubes in steam after the accident. Japanese study was inaugurated in 1960 by the Fuel Safety Specialists' Committee (called NEN-ANSEN), and conducted by myself; internal pressure burst tests on stainless steel and zircaloy cladding tubes were carried out. The 20 years' activity of this committee has been overviewed in detail in the literature published by NEN-ANSEN itself[1] as well as my article[2] in *Atoms in Japan*, and so I do not want to repeat it here. Those who are interested are asked to refer to those publications.

An experimental study on high-temperature oxidation of a zircaloy tube in steam was carried out by Kawasaki of the Japan Atomic Energy Research Institute (JAERI). It showed that thick but porous oxide film is formed on the internal surface when steam flow is stagnant, with less consumption of metal than on the external surface, but that a larger amount of hydrogen is diffused into zircaloy; the heaviest hydriding occurs at a certain distance from the burst position in both the upper and lower axial direction of the cladding tube when the clad temperature exceeds, say, 1,150 °C. This may cause considerable embrittlement. Peak Clad Temperature (PCT) in current LOCA analysis is based on the calculated value computed by the code which gives a considerably higher temperature value and it is to be compared with the critical PCT value of 1,200 °C. Therefore, it is my judgement that the present ECCS criteria of 1,200 °C, 15%, in Japan is still reasonable.

It is easily predicted from the experimental data hitherto obtained that a considerable number of fuel rods will be destroyed in case of an accident like TMI. It seems, however, that there is no urgent need to verify this through experiment. An unpredicted accident due to another cause may occur, but the amount of fission product release is thought not to exceed, in any case, the scale used in the largest hypothetical accident assumed in the present hazard analysis. No additional experimental study has yet been recommended in Japan from the regulatory standpoint after TMI. Discussion from various points of view, is necessary, before experiments on so-called severe core damage will be added to safety-oriented study.

5. Fuel Safety Study Today

As more information is collected on fuel behavior in a major accident, fuel safety study should be directed toward fuel failure in medium or small-scale accidents as well as in slight departures from transience, which are more likely to occur. Recommendations on the construction of a special reactor for fuel behavior in PCM (FSTR) as well as tests in NSRR of irradiated fuel from the power reactor have been made under this concept by the Planning Committee on Fuel Safety Research of the Japanese AEC, conducted by the author.

In 1978 I proposed the necessity of development of high-duty fuel to be fitted to the modified mode of LWR operation which is expected to be required in the coming 20 years in Japan. This development program will be realized from this year by MITI. Development of PCI-resistant fuel, as well as fuel fitted to the elongated burn-up operation expected in the advanced mode of refueling program will be included, and a safety study on the fuel developed by this method will also be carried out.

After more than 10 years of experience with power reactor fuel, a tremendous amount of data to be used for statistical analysis has been obtained in Japan, since examination of the total number of fuel assemblies has been done in the past. The achievement of the world's smallest percentage of leaks in LWR fuel in Japan is due to not only higher quality control and workmanship but also the adoption of various regulatory recommendations mentioned above. The development of high-duty fuel and the study of its safety will be carried out based on this experience.

A examination facility for the power reactor fuel assembly actually used was proposed by myself 10 years ago; I called it a "Pathological Anatomy Facility for Power Reactor Fuel." This has been budgeted and realized by the JAERI in Tokai, and has been operated since the end of 1979. Post-Irradiation Examination (PIE) on 15×15-type PWR fuel from

Mihama No. 3 reactor as well as on 7×7-type BWR fuel from the Tsuruga reactor which achieved burnup of 22,000 MWD/T has been inaugurated. The results will be presented at the Specialists Meeting of the IAEA, International Working Group for Fuel Performance and Technology, to be held in November next year in Tokyo, organized by us.

6. Our Contributions

In the above-mentioned achievements in LWR fuels in Japan for the past years as well as in the study of their safety, my laboratory of fuel technology at the University of Tokyo has been in service as the nucleus of activity. I would like to continue such contributions also in the future.

Description of the fuel development and fuel safety study in newer type power reactors, such as ATR, the multipurpose high temperature gas-cooled reactor (VHTR), and the fast breeder reactor, will be given elsewhere. I have been participating in these programs during the past 15 years and would like to continue contribution as well in the future.

References

1. Nuclear Safety Research Association: NEN-ANSEN—History and Activity, NEN-ANSEN PE-1 (1980).
2. Y. Mishima: *Atoms in Japan*, **24**, 7–11 (1980).

Discussion: Part III

Dr. BENEDICT

Professor Pigford. can Three Mile Island #2 be restored to service?

Dr. PIGFORD

The answer to the question of whether there is a possibility is "yes". But obviously there are barriers to doing it. The first barrier, of course, is the cleanup. The clean-up begins with removing large amounts of radioactive matter like 800,000 Ci of cesium, strontium, and contaminants. However, it seems to me that the only real barriers are the licensing and, of course, the public acceptance. But I think these will be overcome. After that, another question is what is the reactor condition. I foresee no problem in removing the damaged core because some of the operation has been done before. In my estimate, the question is what is the condition of the pressure vessel, because it was subjected to probably appreciable temperatures. To me, the largest question is the technical question: Can the pressure vessel be reused? It is hard to know with some feeling of probability.

There is another barrier, the financial barrier. A serious determination to survive Three Mile Island will cost the company a lot of money. Dr. Davis mentioned earlier today that perhaps some federal support in accomplishing the cleanup may be necessary. It is my understanding that the United States government and the company are spending a lot of money. To my knowledge the main cost to the company is the cost of replacement of electricity. But the total cost is very very large.

So the financial barrier is another one. Setting aside the question of who pays the money, I think, eventually, we will come to the question of the condition of the pressure vessel. If that is all right, personally I see no technical barrier to restoring the plant and continuing the operation.

Dr. DAVIS

Can Three Mile Island #2 be cleaned up and put back into operation? I hope "yes" since Bechtel has a contract with the company to do the job. I agree with Prof. Pigford. There are concerns about the reactor vessel. Probably the head must be replaced, but not the vessel itself. Also there are concerns about the embedments in concrete. However, recent entry showed activity levels to be lower than anticipated, the equipment in reasonable shape, and walls lower in activity than floors, which indicates that activity was washed (and can be washed) off the walls.

The plant must be cleaned up, and the extra cost beyond that will probably be much less than for a new plant. The company has a severe financial problem due to the cost of replacement power and lack of rate relief. Some federal action may be needed. Costs will be high based on our "guesstimates." A problem is where the money will come from. Costs are being increased greatly by delay of local and federal regulatory actions (or inactions), and there is also increased cost to the company of keeping Three Mile Island #1 out of service—though it was ready to operate again after a refueling shutdown.

Part IV

Transition to Fast Breeder Reactors

A Breeder Strategy to Solve Uncertain Future Problems in Energy Supply: The French Example

Jean F. PETIT

1. Introduction

The French energy situation at the present time, in terms of resources and of dependence on foreign supplies, as well as in terms of political uncertainties and financial costs, is such that an extensive program of electricity generation from nuclear power is a vital need. Such a program was initiated five years ago. At the present time about 11,000 MWe are installed and in operation (about 20 plants) and 30,000 MWe are under construction. These plants are mainly PWRs. This should lead in 1990 to a situation where 60% of our electricity will be produced by nuclear power, corresponding roughly to 30% of our total energy needs. Our dependence on foreign supplies would then be decreased to 55%, while it is still 75% in 1980.

In the French context a nuclear strategy only based on thermal reactors like PWRs is not acceptable for the following reasons:

First, because the corresponding uranium needs (10,000 t/year between 1995 and 2000) would exceed our national and foreign resources as early as 1995;

Second, because the large plutonium production would rapidly lead to a difficult problem, either in terms of use if the fuel is reprocessed or in terms of waste if it is not.

Fast reactors, in contrast, allow a much better use of the energetic potential of natural uranium (50 to 70 times higher), which can be almost completely burnt through plutonium transformation. Also the very large amounts of depleted uranium coming from our enrichment plants operation can be used. The problem of resources is thus completely solved. In addition, fast reactors allow us to use the plutonium coming from water reactors after fuel reprocessing and to control easily the balance of plutonium either by producing new plutonium (breeding operation) or by burning it (fertile fuel replaced by reflectors).

Coordinator for Breeder Reactors, Commissariat a l'Energie Atomique, France.

The French strategy is in consequence also based on the intensive development of fast breeder plants which could produce up to 10 to 15% of nuclear electricity at the end of the century.

2. French FBR Development Program

2.1   Introduction

This ambition is supported by a coherent but cautious development program which was initiated in 1950 and in which each step was decided only on the basis of the results of the preceding one.

Rapsodie, the first experimental reactor, began operation in 1967 after more than 10 years of main-component full-scale tests. Its in-pile experience on fuel and components were immediately used for the design and the out-of-pile testing of the components of the Phenix reactor, the prototype demonstration reactor for electricity production which went into operation in 1974, while Rapsodie was then oriented towards intensive and statistical fuel testing and experimentation. Super Phenix, the nuclear steam supply system (NSSS) of the large power prototype plant of Creys-Malville, was ordered in 1977 when it was possible to take benefit of 3 years of Phenix operation and 10 years of Rapsodie operation. After Creys-Malville, which should come in operation in 1983, EDF, our national electricity utility, intends to order in 1985 a small series of identical power plants as close as possible to a commercial series. A NSSS called Super Phenix-II is under preliminary design for this project. The aim, in general technological continuity with Super Phenix-I (Creys-Malville Plant), is to get a simpler and cheaper reactor.

Of course, fuel cycle installations (fabrication and reprocessing plants) have followed the same evolution, though beginning later for reprocessing.

This R & D and construction program is based on constant technological options which have up to now been entirely confirmed in terms of validity, reliability, and safety. The main ones are:

Mixed oxide ($UO_2 - PuO_2$) for fuel;

Liquid sodium as a coolant;

A pool concept for the primary system organization, that is, primary pumps and heat exchangers contained entirely inside the primary vessel from which no primary sodium goes out (the other possible concept being the loop one where pumps and IHXs are outside the vessel, which requires external primary pipes).

I would like to give now some more details on the French reactors themselves.

2.2 Rapsodie

Rapsodie, an experimental reactor of the Joyo type located at Cadarache Nuclear Center in the south of France, went into operation in January 1967. Its initial thermal power (no electricity production) was 24 MW. It was upgraded to 40 MW in 1970. In 13 years of operation the reactor has given a lot of information and experience. Its objectives must be considered as fully obtained: to demonstrate the ability of a liquid sodium-cooled fast reactor to operate safely and reliably and to qualify fuel at high burn-up. Only three figures are sufficient to illustrate the success in this regard: more than 28,000 fuel pins have been irradiated with a nominal burn-up of 11%, some experimental pins having reached up to 24%. However, Rapsodie at the present time operates only at around 24 MW power because of the detection of sodium aerosols inside the inert safety envelope surrounding the primary system. In spite of long and detailed investigations, the possible leak is so small that it has not been located yet. At 24 MW, no signal has ever been detected.

2.3 Phenix

Phenix, a demonstration power plant of 250 MWe situated in Marcoule (the south of France), was commissioned at full power in July 1974, less than 6 years after the beginning of construction. Since it has produced more than 800 million KWh, which corresponds to an availability factor of 65%. During the last two years, the availability factor was higher than 90% and the load factor higher than 80% (the load factor is representative of the time during which the reactor produces electricity at full power, while the availability factor is related to the time during which the reactor is available either for electricity production or for normal operation done at shutdown like refueling or planned maintenance). More than 73,000 fuel pins have been irradiated to their nominal burn-up, which is now quite higher than initially planned (66,000 and 82,000 MWD/t (owing to two zones of the core) compared to 50,000) with only a single significant fuel pin defect which was immediately detected, the consequence of it having been a two-day shutdown to discharge the corresponding fuel element. As far as safety is concerned, neither accident nor significant incident has occurred, and the radiological doses to the personnel as well as the liquid and gaseous radiological releases are negligible (several hundred times lower than the acceptable limits which themselves are fixed in relation to exposures of any human being on the earth coming from cosmic rays and natural radioactivity). Maintenance is also a very important aspect. In 1976 and 1977, several periods of shutdown or operation at 2/3 of the nominal power were necessary to cope with a common defect affecting IHX's upper part (abnormal deformations due to insufficient design for thermal transients account, leading in two cases to a limited secondary sodium leak). Over 18 months, the six IHXs were successively taken out of the main vessel, washed, decontaminated, disassembled, repaired, and reinstalled, so that in 18 months the full nominal power could be reached again. The cost of the whole operation in terms of man-rem was very low (less than 30 man-rem) while the annual mean dose for 5 years of operation (1974–79) is less than 25 mrem per person. Since then, as I mentioned, the load factor is higher than 80%, while this factor for the reactor's whole life up to now is 55%. The maintenance ability of a pool-type reactor has thus been fully demonstrated. On all aspects, including design options confirmation, reliability, safety, and maintenance ability Phenix's record is up to now remarkably positive.

2.4   Super Phenix

Super Phenix represents the last step prior to the industrial development of fast reactors in France. It is a 1,200 MWe plant being erected at Creys-Malville along the Rhone River in the southeast of France. The construction began in 1977, and the commissioning is expected at the end of 1983. At the present time, the construction schedule is only a few months behind the initial planning. The reactor building is achieved. The different reactor vessels with their internal structures and the irradiated fuel storage tanks, all having been assembled in a large workshop on the site, were just installed inside the reactor building. The workshop will now be used for assembling large components like IHXs, sodium pumps, and S.G.

2.5   Fuel Cycle Facilities

A breeder program also implies the parallel development of the corresponding fuel cycle installations. Fuel fabrication does not raise too many problems. Rapsodie, Phenix, and Super Phenix's supply is easily assured by CEA present installation (At-Pu at Cadarache). Reprocessing is more complicated (mechanical dismantling of fuel assemblies, high concentration in radioactive fission products and plutonium, final conditioning of plutonium, criticality problems all along the chain). France already has experience with pilot or experimental installations since several tons of mixed oxide at high burn-up (higher than 100,000 MWD/t for some Rapsodie fuel and higher than 75,000 MWD/t for Phenix fuel) have already been fully reprocessed. The Rapsodie cycle has been closed for several years. More than three cores have been treated, mainly at La Hague-ATI. The Phenix cycle was closed in 1979 (La Hague-UP2 in dilution with gas reactor fuel and Marcoule-SAP). A 5 ton/year capacity demonstration plant is being erected at Marcoule and should be commissioned at the beginning of 1983. Its capacity allows the reprocessing of the fuel equivalent of 2.5 reactors like Phenix.

2.6   Future Plans

Super Phenix is a prototype. It will produce electricity at a price comparable to a present modern conventional plant burning desulfurized imported fuel under French conditions. The final stage of industrial development which has recently been entered aims at better competitiveness with present PWR electricity (more precisely, the objective in term of KWh cost is 15 to 20 % higher than PWR). As far as reactors are concerned, Super Phenix design is used as a reference to be simplified and optimized for Super Phenix II, a NSSS under preliminary study by Novatome which could equip two to four breeder plants that might be ordered by EDF in 1985 or 1986. Investment savings are foreseen with an increase of power (1450 MWe) without modifying the vessel's size, a series effect (mainly on components), a site effect (four reactors on the same site), and design simplifications on vessels (roof, dome) and on the secondary system.

For the fuel cycle, two large plants (TOR for fabrication and PURR for reprocessing) are under preliminary design by CEA and its subsidiary COGEMA. Their capacity is consistent with the needs of the reactors here mentioned above. PURR in particular (100 to 150 ton/year) will be also devoted to Creys-Malville reprocessing. The schedule is such that a decision on order is to be taken at the same time as that on the reactors in order to achieve coherent commissioning of all plants, starting around 1992. Co-location of the fuel cycle installations and of the four reactors on the same site is taken into account as a hypothesis in order to largely decrease fuel transportation and safeguards problems.

2.7   International Cooperation

The program that has been outlined is deeply involved with European cooperation. At the R & D level and at the industrial level, France and Italy first, then France and DeBeNe (FRG, Belgium, and Holland) have signed agreements integrating R & D programs on the one hand and granting the French system license to their partners on the other hand. The five countries of the continental European community are thus engaged in a common venture to develop, realize in their own countries, and propose to others, fast reactors on common designs. Besides, as far as utilities are concerned, Super Phenix has been ordered by an international association called NERSA, between EDF (51%), ENEL for Italy (33%), and RWE for FRG (16%).

Greater involvement with other countries outside Europe in the future would be in my opinion of great benefit for all parties.

3. Conclusion

In conclusion, I want to stress the following point. A country like France has obviously no other choice for ensuring an energy supply, during the uncertain next 30 or 40 years at least, than developing nuclear energy from thermal and fast reactors. If we scientists and engineers take the technical responsibility for realizing the equipment program, it remains necessary that the corresponding short- and long-term political decisions be taken to ensure political, financial, technological, and public assistance responsibility for maintaining them efficiently over a long period of time. We have the luck in France to be at the present time in such a favorable situation. Some other friendly democratic countries are not. I do deeply hope their situation will change as soon as possible, because one cannot go alone today in the nuclear venture on a purely single-national basis. It is the world energy supply, and hence the world's immediate future, which is in question.

RDD & D on the Fast Breeder Reactor

Shigehiro AN

The title of my talk is RDD & D on fast breeder reactors (FBRs). RDD & D might be quite a new acronym. It stands for research, development, demonstration, and deployment. Deployment, which is the final stage of RDD & D activity, is the process of commercialization of FBR technology, including the establishment of a necessary framework for adapting it in our society. I would like to discuss the major tasks to be accomplished in the process of FBR demonstration and deployment.

I. FBR RDD & D Activities in the World

First of all, I would like to review very briefly FBR RDD & D activities in the world. Figure 1 shows the FBR development schedules in the world. Let me go over this figure quickly. In the U.S.A., EBR II has been in operation successfully for the last 14 years. FFTF went operational in Feburuary 1980, and since then low-power testing has been going on. The construction of the Clinch River Breeder Reactor (CRBR) was deferred by President Carter in 1977. It is reported, however, that as far as fabrication of its components is concerned, 60 to 70% of it has been already completed. If this project is re-activated under the new administration, it would be a valuable contribution to the world FBR community.

Moving on to the United Kingdom, PFR began operating in March 1974. However, it could not go on to full power smoothly because of leakage in the steam generator. Several years were required to repair this component, and it attained full-power operation in 1977. After that, however, steam generator leakage has occurred again, and the operation is going on and off because of the intermittent occurrence of this steam generator trouble. The feasibility of constructing CDFR, which is the commercial demonstration reactor for the U.K., is now being discussed, and I understand the future of it will be finally determined by 1985.

Nuclear Engineering Research Laboratory, University of Tokyo, Tokyo, Japan.

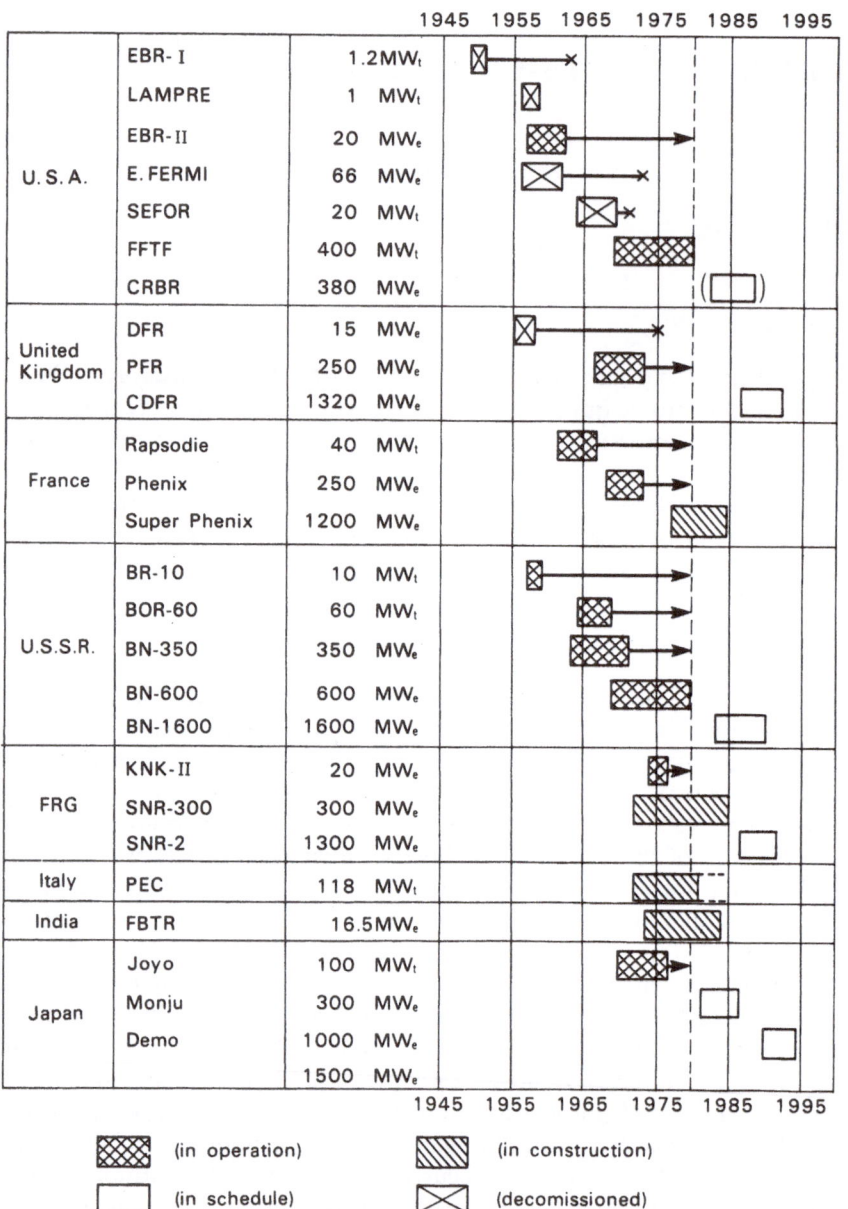

Fig. 1. FBR development schedules in the world.

I would like to forego an explanation of the French situation since it was described excellently by Mr. Petit of CEA. In the case of the U.S.S.R., BN-350 became operational in November 1972. This plant is unique in its dual purpose design: it is used not only for generating electric power but also for desalting sea water. The next plant, BN-600, has also started sending electric power at 30% of the full power level already. Since BN-600 is now in operation, the design of BN-1200 is in the stage of detailed design for construction. It will come onstream in 1990 at the earliest. Germany is now operating an experimental fast reactor KNK-II. Her prototype reactor SNR-300 has been under construction for quite a long time and I hope it will be completed in several years. Italy has an experimental reactor called PEC which was scheduled to begin operations in 1979, but the schedule is far behind at present because of delays in the licensing process. In India, construction of FBTR is continuing. It was scheduled for completion in 1979 initially, but apparently the schedule will be delayed until 1983.

In Japan Joyo began operating in April 1977 and reached 50 MW (thermal) in July 1978 and attained 75 MW (thermal) in July 1979. So far this reactor has been operated for about 10,000 hours, and the burn-up of fuel is at the level of about 27,000 MWD/T. There is a plan to increase its thermal output to 100 MW by modifying the present core to the Mark II core. Mark II core will begin operating in 1983. As for the prototype reactor Monju, the safety evaluation process for it is going on and the plant is expected to be operating in 1987. PNC and a utility group have finished several design studies for a demonstration reactor, and construction of it will be begun when the Monju reactor is completed.

In summary, except for France and the U.S.S.R. who are going smoothly into the FBR demonstration stage, most countries are not so successful in implementing their original FBR RDD & D programs at present. But every countries still think that FBR should be the main reactor necessary in near future and they are expending considerable efforts on it with perseverance.

II. Issues Important for FBR Demonstration and Deployment

1. General

What are the major obstacles which have hindered national FBR programs in various countries? They are political, societal, administrative, technical, etc. This diversity of problem areas is characteristic of the modern society. It does mean that the technical issues of FBR do not belong to just one independent small group of issues. In other words, technical issues are closely related to those various issues in diverse disciplines. Therefore it is essential for us engineers to endeavor not only to solve technical problems but also to contribute to the solution of problems in other disciplines from the technical viewpoint.

Now let us list the major technical goals for FBR commercialization based on this understanding. These are (1) scale-up of plant capacity for commercialization, (2) establishment of practices for design, construction, and operation of highly safe and reliable plants, and (3) establishment of FBR economy, including FBR fuel cycle.

2. Plant Scale-up

The goal of FBR RD&D is the realization of plant scale appropriate for FBR commercialization. It has been generally accepted that the capacity of commerical FBR will be 1,500–2,000 MWe from economical viewpoint. However, we have to study the technical risk of various scale-up schemes from 250 MWe prototype plant and necessary R&D resources for overcoming it. In other words, we should try to find the best compromise between technical risk and economy in determining the optimum plant scale for commercialization.

3. Safety

3.1   General Issues

In establishing sound and competitive FBR technology, it is important to establish a methodology for design, construction, and operation of highly safe and reliable plants. For this goal, it is necessary to prepare reasonable safety codes and standards for FBR manufacturers. These should be prepared by not only a licensing body but also industry and utility groups. At this moment, there is a general understanding that the defense-in-depth principle is a basic philosophy for every safety aspect of design, construction, and operation. However, this principle alone is not sufficient since it is a kind of one-way guiding principle for enhancing safety and does not answer the question of how a defense is deep enough. It is clear that factors other than technology should be taken into consideration, and it is important to pursue a worldwide consensus on the principle of FBR safety in this respect. Specifically this consensus should include understanding of the following items:

(i)   What is ther equired level of reliability of a decay heat removal system and of a reactor shutdown system, and how should this be confirmed?

(ii)   To what extent should coolant pipe rupture be considered in the safety design? Could we rely on the leak-before-break concept?

(iii)   To what extent should a core-disruptive accident (CDA) be considered in the safety design?

(iv) To what extent should local malfunction be monitored and diagnosed?

3.2   Core-Disruptive Accidents (CDA)

It is generally accepted that a CDA is quite an unlikely event. However, if it should happen in a reactor without proper safeguards against this type of accident, the consequences might be serious. Therefore it is necessary to evaluate the consequences of this type of accident, and if the result is unacceptable even considering its low occurrence probability, proper design margins should be included in the design so as to reduce the consequences to the acceptable level. This is the way both the USNRC and French regulatory authority have managed the CDA issue in considering the safety of proposed reactors.

Another important subject is the clarification of the phenomena involved in CDA. Significant efforts have been devoted to obtaining a clear understanding of initiation, transition, dispersion, and post-accident heat removal phases of CDA. The energetics aspect of CDA has been a key issue in this research areas for 15 years. Recently, sodium fires and melt-through and resultant hydrogen release in the reactor cavity during the cooling of molten fuel debris have also become important phenomena to be understood in considering safety provisions for containing CDA. We are sure that it will be possible to predict the degree of risk of CDA before FBR commercialization, although we have to evaluate a long list of phenomena (including the above ones) in the safety discussion. In the meantime, however, designers may well be advised to build into their design the capability to mitigate such events until their insignificance can be confirmed.

3.3   Technical Options for Enhancing Inherent Safety

Several options are proposed to enhance the inherent safety of FBRs so as not to be required to consider CDA as a design-based event. One of them is the adoption of a heterogeneous core to reduce the sodium void coefficient. It has been generally accepted that the reduction of a positive sodium void coefficient is one of the major strategies for increasing the inherent safety of FBRs. When the core size becomes larger, however, it is necessary to build in or add some features to the core design even if it is required to keep this coefficient at the same level as the present prototype FBR since as the size of core increases this coefficient becomes more positive. The heterogeneous core is the core which includes blanket regions like islands or radial zones. The advantages of this heterogeneous core are that (1) this core has a low sodium void coefficient of 0.6–0.7% $\triangle$ k/k, as compared to about 2% $\triangle$ k/k in the case of homogeneous core, and (2) a breeding ratio of 1.35 to 1.4 can be obtained by this configuration. The latter advantage of this configuration is reduced by a 20–30% increase in fuel inventory due to the dilution of the core with blanket material over the inventory of homogeneous cores. Thus it should be noted that, even in the case of larger fuel pellet diameter, only a slight reduction in fuel doubling time, which is the key parameter determining FBR fuel cycle cost, is expected with the adoption of this configuration.

The former advantage depends upon the high incoherency of sodium boiling initiation between core and inner blanket regions under accident conditions. Therefore, it may be required, in order to capitalize on this advantage, to obtain the assurance of this incoherency with sufficient reliability, in addition to experimental confirmation of this reduction in the sodium void coefficient.

4. Assurance of Reliability

High reliability of the plant is an essential goal for commercialization. From past experience, it is clear that the reliability of the steam generator is one of the focal points. It is extremely important to establish design rules and construction procedures for obtaining a highly reliable steam generator system. Assurance of the maintainability of the system, particularly of those components which are in contact with liquid sodium, is a great contributor to the goal. Past experience in the deployment of nuclear reactors in our society indicates that it is essential to assure adequate maintainability of major components in the plant. The design should be reviewed deliberately from this viewpoint.

Reliability and maintainability are dependent also on the configuration of the plant. There have been various discussions on the system configurations of FBR. The question of whether pool type or loop type is better is one of the classical ones. It is needless to say that each has its own advantages and disadvantages. It is therefore extremely difficult to make any conclusive remarks on this question. What I would like to stress on this occasion is that the selection should reflect the philosophy of the designers involved and that it should be consistent with the design philosophy of the structure, the fuel handling scheme, the steam cycle, etc., for realizing the design goal of high reliability and maintainability.

5. Fuel Doubling Time

It is generally recognized that the breeding ratio of FBRs today is not large enough to cope with the pace of moderately expanding electric power demand if the power supply system is composed of a FBR without any supply of plutonium from outside of the system. Therefore I am sure it is necessary to reduce the fuel doubling time. Several directions have been discussed in the FBR community to improve performance of the FBR core. Among them are the improvement of present-day oxide cores, employment of the heterogeneous core concept, and employment of advanced fuels. If the first option is used, fine tuning of the design parameters is necessary to shorten the doubling time and therefore reduce the fuel cycle cost. For example, fuel pin diameter should be increased from the present value of 6–7mm to 8–9mm; smeared density of fuel should be increased to 88–90% of theoretical density; cladding thickness should be reduced, but the capability to withstand burn-up of above 10% should be included; linear power rating of 500 W/cm should be realized; etc. It should be noted that considerable research and development is necessary for realizing each of these proposals and filling the gap between the breeding ratio of the present-day FBR and the target value compatible with the expanding future role of FBR, which is 0.1–0.2.

Use of carbide fuel or metal fuel is another possibility for improving the economy of FBR. These fuels have higher fuel density and higher thermal conductivity than oxide fuel and therefore have the potential to improve fuel doubling time. It is needless to say, however, that it takes considerable time and resources to develop a new fuel technology. Before going in or in parallel with this direction, it would be worthwhile to develop advanced cladding materials by which the thicknesses of both oxide fuel cladding and fuel assembly ducts may be reduced, since recent increases in these thicknesses are one of the major factors contributing to the reduction of the core breeding ratio.

Let me add one thing before going on to the next subject since I would not like to leave the impression that the activity for this goal has the primary weight. From my point of view, the primary weight should be placed upon the smooth introduction of FBR with small plutonium inventories. By doing so we will have sufficient time for exploring various options to improve FBR fuel doubling time in parallel with the learning process necessary for the introduction of FBR in our society. In this respect, the reduction of fuel doubling time should not be a near-term goal, though it is a long-term technical necessity for FBR deployment.

Finally it should be noted that fuel doubling time is closely related to the capability of the fuel cycle process system available. This fact suggests to us that, in the first place, it is necessary to demonstrate the technology of FBR fuel reprocessing and waste management in parallel with the demonstration of FBR power plants, and that, secondly, it is important to have a well-coordinated program of FBR introduction into the now established LWR nuclear energy supply system, which should include deliberate consideration of the long-term behavior of an FBR-LWR combined fuel cycle system.

6. Economy

The economic competitiveness of FBR is a well discussed but still rather speculative issue. In any case it is a final goal of any RDD & D activity. Electricity generation cost consists of capital cost, operation and maintenance cost, and fuel cycle cost. It has been generally accepted that the capital cost of FBR is higher than that of LWR but that the fuel cycle cost of FBR will become considerably less than that of LWR when the price of natural uranium eventually rises in the future. It is difficult to come up with an accurate estimate of the capital cost ratio between commercial-scale FBR in the introduction phase and LWR at present. There are several studies which indicate the present situation to a certain degree. Values of 1.2 to 1.5 are claimed by researchers in the U.S.A., 1.3 to 1.4 by the French, 1.5 by the U.K., and 1.3 to 1.4 by the U.S.S.R. As for the price of natural uranium, the supply-demand relation in the future market is the primary element of uncertainty. The prediction of the future market price of uranium, therefore, has been one of the controversial issues in the INFCE study, as you well know. The important implication of the output of the study is, however, that if nuclear energy is expected to supply a considerable fraction of energy in the future, the uranium market will become tight eventually and the FBR will be the vital replacement for the LWR. Reprocessing cost is another important component of fuel cycle cost. As the reprocessing of FBR fuel is in its infancy, it is difficult to predict its cost precisely. The present cost lies around several ten percent of capital cost. It is necessary to evaluate whether this should be used in calculations for the future or not, since it is a considerably higher level than was once expected. The last but not the least important factor to be considered in the discussion of FBR economy is the FBR introduction scheme. Apparently it is rather difficult to establish the economy of the FBR if it is introduced intermittently. FBR industry in its infancy is necessarily not so strong. Therefore it is vitally necessary for both utility and industry to agree, roughly at least, on an FBR introduction program.

7. Organizations for Demonstration and Deployment

It is clear that the dominant actor in the commercialization of FBRs is the utility. In our case, the utility should establish an organization to determine the order of demonstration plants, after the construction of

Monju starts. Corresponding to this action on the utility side, the manufacturing industry may be required to establish and/or strengthen its organization with sufficient engineering capability to integrate the total plant. The result of R & D in the national sector should be transferred to this organization as smoothly as possible. This organization should be continued at least until the learning process is finished on the first demonstration plants and competitive technology has been established within the manufacturing industry.

As mentioned in the section on safety issues, there will remain several areas needing further R & D even after the demonstration reactor is completed. This situation is easily imagined if you look at the active safety R & D for LWRs in the world. These R & D projects will be done by manufacturers and utilities as well as governments, and are one of the best items for international cooperation. The important issue here is how to initiate and coordinate this R & D in parallel with keeping a competitive market situation.

International cooperation in the deployment stage is one of the most important tasks in organizing the deployment process. There may be many possible schemes of cooperation; the most popular form of cooperation is information exchange; the next stage may be technical cooperation in the R & D area. These are the same as those now existing, such as the following:

a) Agreement between the U.S. DOE and PNC of Japan in the field of LMFBR;

b) Agreement on technical cooperation in the field of LMFBR between KFK-Interatom, CEA, and PNC;

c) Agreement on technical cooperation in the field of LMFBR between JAERI, PNC, and UKAEA;

d) Agreement on scientific and technical cooperation between the U.S.S.R. and Japan.

Where commercial interests are involved, however, it is generally recognized that it is difficult to organize a cooperative scheme. We can see and hear that excellent cooperation on FBR is now working in the European Community. From the viewpoint of the Japanese nuclear industry, which has precious experience in the process of introduction of foreign technology, however, it is vitally necessary to have our own technology, or at least to have foreign technology modified such that it is harmonious with Japanese engineering practice as well as social and environmental conditions. It will apparently be necessary for us therefore to establish our own system concept, even when we seek the benefit of international cooperation.

III. Conclusion

The FBR is accepted as a central technology for the best utilization of nuclear energy in the future. Various issues remain to be resolved in the process of FBR demonstration and deployment. These not only are technological and economical in nature but also involve social and international policy. There are, however, good signs indicating that these are solvable effectively and in time through the efforts of each concerned national and international sector. I am sure this process can be further accelerated if effective international cooperation is promoted. Our Department of Nuclear Engineering at the University of Tokyo has devoted its research efforts to R & D for FBR for the last 20 years, and we can and will continue to do so in the future for the benefit of humanity.

Discussion: Part IV

Dr. OYAMA

Thank you very much, Dr. Petit and Prof. An. Dr. Petit has presented the consistent French approach to the uncertain energy future and Prof. An has presented a very organized review of the questions the FBR community is confronted with. I wonder if Dr. Petit would like to comment on Prof. An's presentation.

Dr. PETIT

I have not much to add to the exhaustive review given by Prof. An. The only thing I want to add is to note the remarkable similarity between the strategies for breeders in France and Japan which are so separated. This partly proves that the breeder must be a necessarily legitimate step in solving worldwide energy supply problems in the uncertain future of the world. Indeed we have the same problem of scarce energy resources in both countries. We chose, therefore, the LMFBR as a part of our nuclear program for surviving even in an energy crunch, and we are confident that it will continue to be a good choice.

Part V

Research and Development of Fusion Technologies

Nuclear Fusion Research: Status and Prospects

Hugh A. B. BODIN

1. Introduction

It is appropriate, on this 20th anniversary of the Department of Nuclear Engineering at the University of Tokyo, to review progress in nuclear science and technology over the last 20 years and to consider what developments are likely in the next two decades. The purpose of the present paper is to address this question in the field of nuclear fusion research, placing emphasis on studies in laboratories outside Japan. Japanese work will be discussed in the accompanying paper by T. Uchida.[1]

Fusion, in contrast to fission, is a young subject. Indeed, 20 years ago it had just become established as a major branch of applied science, together with plasma physics, as a new field of basic research. Fusion is still essentially in the research stage, although fusion reactor technology is now a rapidly expanding field. Scientific feasibility—the realization of a net energy gain from the hot plasma—has yet to be demonstrated. However, machines are now being constructed that will provide conditions which come very close to or may even reach those required to achieve this major milestone. By comparing the development of fusion with that of fission, it is clear that a commercial fusion power plant is unlikely to be built until the next century. However, by the year 2000 answers should be obtained to three crucial questions: (1) Is fusion scientifically feasible? (2) If so, can an economically feasible reactor be built? (3) If it can, what is the time scale for the introduction of fusion power?

The basic principles of fusion research are described in Section 2. In order to produce useful fusion power, it is necessary to satisfy the Lawson criteria, which for the deuterium-tritium reaction are:

$$T \gtrsim 10 \text{keV} \ (10^8 \ {}^\circ\text{K})$$

$$n\tau > 10^{14} \text{cm}^{-3}\text{s}$$

where T is the temperature, n the number density (ions/cc), and τ the con-

UKAEA, Culham Laboratory, Abingdon, Oxfordshire, U.K.

145

finement time (in seconds). A plasma can be confined by gravity, as in the sun and stars, by a magnetic field since it has high electrical conductivity, or by its own inertia. There are many magnetic field configurations which can in principle confine plasma and provide the necessary thermal insulation from material walls, both in linear geometry and in toroidal geometry with closed field lines. Research is now converging on a few of the most promising systems, mostly of the toroidal type. In inertial confinement a small fuel pellet is compressed and heated very rapidly by laser light or charged particle beams, so that fusion reactions occur before it has time to fly apart. In practice a series of micro-explosions is proposed. This work developed in the 1970s, much later than magnetic fusion. Nevertheless, machines which might in principle produce a net energy gain from the pellets are being designed.

Fusion reactors are described in Section 3. In principle fusion can provide an unlimited source of energy which would meet the needs of mankind indefinitely, since the basic fuel can be extracted from water and is cheap and plentiful. However, as presently conceived, a thermonuclear reactor is very complex and fusion power is likely to be expensive. From an environmental viewpoint, fusion is an attractive energy source. Many conceptual fusion reactor designs have been proposed, and there are now some quite detailed studies.

In Section 4 magnetic fusion research is reviewed. There has been steady progress over the last 20 years, which has become rapid recently, in obtaining improved parameters and a better understanding of plasmas. In one system—the tokamak—conditions approaching those needed for a reactor have been demonstrated using high-powered neutral injection to heat the plasma with temperatures up to 7 keV ($\sim 7 \times 10^7$ °K) accompanied by intense thermonuclear neutron emission. Values of $n\tau \sim 3 \times 10^{13}$ cm^{-3}s have been reached at lower temperatures. However, the confinement time still falls short of what is required to obtain a net energy gain by a factor of 5 or 10. Progress in some of the most promising alternative (non-tokamak) lines is briefly described. Although the tokamak is presently the most advanced system experimentally, it has not yet been established that it will lead to an economic fusion reactor nor, if it does, that it will be the best system for development as a fusion power plant. It is, therefore, important to continue studies on some other magnetic configurations.

In inertial confinement (Section 5) there has been rapid progress in the development of large lasers. For example, lasers with an energy $\gtrsim$ 10kJ have been used to compress small fuel pellets. Temperatures of $10^7 - 10^8$ °K have been observed with thermonuclear neutron production, but much higher density compressions are still needed. Pellet compression experiments using relativistic electron beams have been reported, and light and heavy ion beam facilities are being developed.

In Section 6 the next generation of large magnetic fusion experiments is discussed, including brief descriptions of three large new machines which are being constructed in which conditions required for a net energy gain might be realized during the 1980s. These are the Joint European Torus (JET) located at Culham in the U.K., TFTR in the U.S.A., and JT-60 in Japan. There is no clear consensus in the magnetic fusion community as to what should follow the machines, but a major step was taken by setting up the international INTOR Study Group, which has made preliminary design studies of a yet larger machine. Large inertial fusion facilities are also planned. The author's conclusions on the present status and possible future developments in fusion research are presented in Section 7.

2. Basic Principles of Nuclear Fusion

2.1 Reactions and Plasma Parameters

The deuterium-tritium (D-T) reaction has the highest cross-section at low energy and releases a large quantity of energy, so it is presently the most favored. It is:

$$D + T \rightarrow {}^4He + n + 17.6MeV.$$

About 80% of the energy released is in 14MeV neutrons which escape from a magnetically confined plasma and are captured in the reactor blanket. Tritium is regenerated in the blanket by the two following tritium breeding reactions:

$$ {}^6Li + n \rightarrow {}^4He + T + 4.8MeV $$

$$ {}^7Li + n \rightarrow {}^4He + T + n - 2.47MeV $$

which completes the fuel cycle in a D-T burning reactor.

It might be possible to use the D-D reactions which eliminate the need for tritium breeding and produce a greater fraction of the reaction energy as charged particles; they have lower cross-sections and require larger values of T and $n\tau$. They are:

$$ D + D \rightarrow {}^3He + n + 3.27MeV $$

$$ D + D \rightarrow T + H + 4.03MeV. $$

It is just conceivable that the p-B^{11} reaction, with an even smaller cross-section, can be used; it has the advantage that virtually no neutrons are produced, so that the reactor will be almost free of the problems of radio-activity.

In order to produce a net power gain, the energy produced from the thermonuclear reactor must exceed the energy required to heat and con-tain the gas, together with that lost from the plasma. There is an un-avoidable minimum energy loss due to bremsstrahlung radiation from the electrons. Since the rate of fusion energy production increases more rapidly with temperature than that of radiation losses, there is a critical tempera-ture, sometimes called the ignition temperature, of about 4.5×10^7 °K for D-T, above which self-sustaining fusion reactions can occur. Once this temperature is exceeded, the charged particle reaction products will, provided they are contained in the system, raise the temperature further to the working value of 10—20keV.

The parameters required for a reactor which satisfy the Lawson critieria (Section 1) are approximately:

magnetic confinement $T > $ 10keV
$$n \sim 10^{14} \text{cm}^{-3} \text{ (dilute gas)}$$
$$\tau \sim 1 - 10\text{s}$$

inertial confinement $T > $ 10keV
(micro-explosion)

$$n \sim 3 \times 10^{25} \text{cm}^{-3} \text{ (superdense solid 1,000} \times$$
$$\tau \sim 3 \times 10^{-11}\text{s} \quad \text{density of liquid D-T).}$$

2.2 The Principles of Magnetic Confinement

The basic force which confines the plasma is:

$$\boldsymbol{F} = \boldsymbol{J} \times \boldsymbol{B} = \nabla\text{p}$$

where $\boldsymbol{B}$ is the magnetic field, $\boldsymbol{J}$ is the current density in the plasma and p = 2nkT (where k is Boltzman's constant) is the thermal pressure. A magnetic field has an equivalent pressure of $B^2/2\mu_0$, and the pressure bal-ance equation may be written:

$$2\text{nkT} = \beta \, B_e^2/2\mu_0$$

where B_e is the field outside the plasma and β is the efficiency factor for the utilization of the magnetic field $(0 < \beta < 1)$. Increasing β in the plasma leads to a reduction in the required field, in the forces on the structure and in the stored energy, i.e. in the reactor cost. For an economic reactor $\beta \gtrsim 5 - 10\%$. A stable plasma in equilibrium which has infinite conductivity would be confined indefinitely in the radial direction by a magnetic field. Because the conductivity is finite, cross-field transport will occur with a perpendicular diffusion coefficient which varies as:

$$D_\perp \propto 1/B^2 T_e^{1/2}$$

corresponding to a confinement time:

$$\tau \propto a^2/D_\perp \sim a^2 B^2 T_e^{1/2}$$

which is very long, *e.g.* $10^2 - 10^3$s, in reactor conditions. In practice the confinement time usually falls considerably short of this ideal classical value (T_e is the temperature of the electrons).

One of the best-known classes of toroidal magnetic field configuration is shown in Figure 1, where the plasma is confined by a combination of a toroidal field, B_ϕ, directed around the torus produced by external coils and a poloidal field, B_θ, due to a current flowing in the plasma which is induced by transformer action. The resulting configuration is helical. This includes the tokamak (Section 4.1) and the reversed field pinch (Section 4.2). In the stellarator system (Figure 2) there is no net plasma current and the required B_θ field component is produced by external helical windings (see Section 4.3).

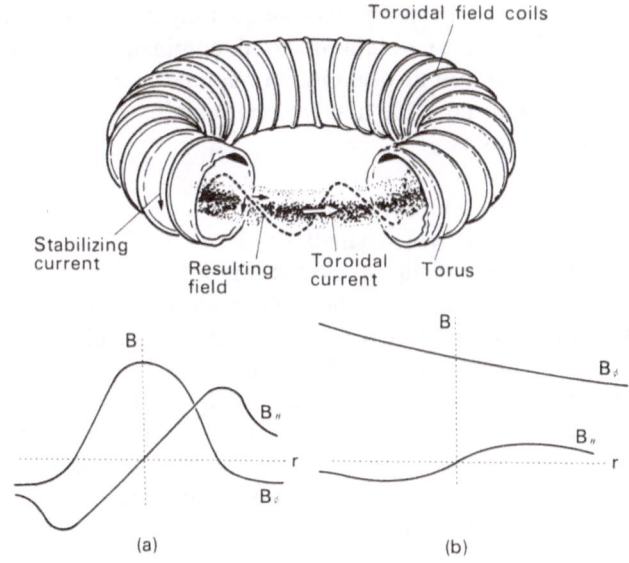

Fig. 1.   Toroidal pinch type systems:
(a) reversed field pinch, (b) tokamak.

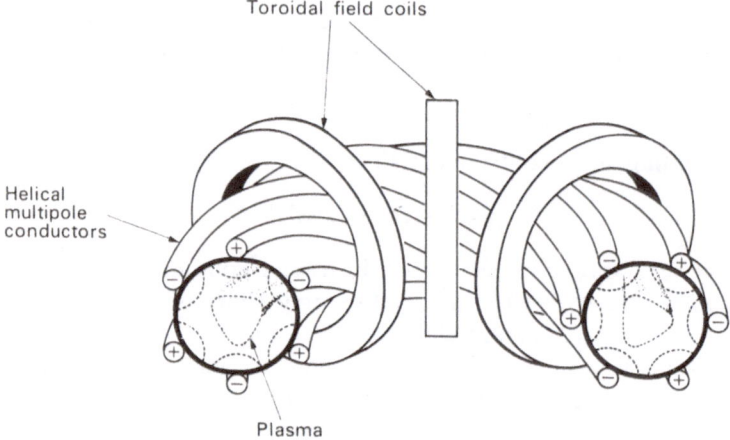

Fig. 2. Stellarator configuration showing helical windings.

A plasma confined in equilibrium by a magnetic field is often unstable to magnetohydrodynamic, fluid-type, instabilities which can cause it to interact with the walls and lose energy. Examples of unstable perturbations are shown in Figure 3. There are also fine-scale micro-instabilities involving wave-particle interactions which can lead to heat loss without large distortions. Whereas 10—15 years ago magnetohydrodynamic instabilities were a major obstacle to obtaining good confinement, in many cases means have now been found whereby they can be suppressed or inhibited. Nevertheless, residual unstable fluctuations remain as one possible explanation for energy losses.

2.3 Principles of Inertial Confinement

The basic idea is to heat a small solid pellet of D-T fuel to fusion temperatures by means of high-powered lasers or charged particle beams (the driver) so rapidly that the pellet will ignite and generate substantial thermonuclear reactions before it has time to fly apart. The fusing material holds together by its own inertia during the reaction time. This idea has been discussed for many years but was thought impracticable because the driver energy was too large ($\sim 10^9 - 10^{12}$J) for a pellet at solid state density. However, the required energy varies as $1/n^2$ where n is the pellet number density and is reduced for a superdense compressed pellet. The energy in a pellet radius, r, is:

$$E_p = 3nkT \cdot \frac{1}{3} \pi r^3 \propto (nr)^3 T/n^2.$$

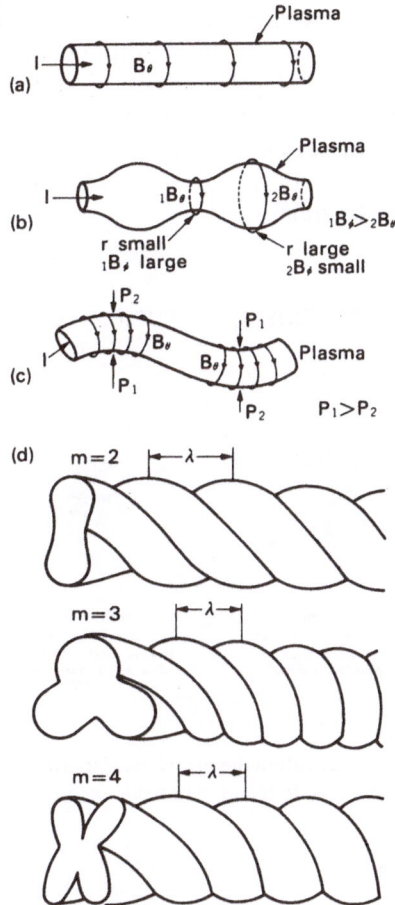

Fig. 3.   Schematic diagram of plasma perturbations due to instabilities which can occur in reversed field pinch and tokamak systems. The top figure shows the stable equilibrium column.

The disassembly time, $\tau \sim r/v_s$ where $v_s = \sqrt{kT/m_i}$, is the sound speed (m_i is the ion mass) so that $n\tau = (nr)/v_s$. Since $n\tau$ and T are fixed by the Lawson criteria:

$$E_p \propto 1/n^2.$$

For example, a density increase of $10^2 - 10^3$ reduces the driver energy to $10^3 - 10^6$J, which is practicable. Furthermore, the laser or charged particle beam can itself compress the pellet core by the required factor. When the pellet is symmetrically illuminated, the outer layers are evaporated and thrown off (ablated) and in so doing impart a compressive force by conservation of momentum due to the rocket effect. This is illustrated in Figure 4. Extensive numerical calculations of pellet compression have confirmed these ideas. A critical factor is the product of the pellet gain (fusion energy out/energy in) and the driver efficiency. For laser fusion, with laser efficiencies $\gtrsim$ 5—10%, pellet gains of 100—1,000 are required, but these appear to be feasible.

INERTIAL CONFINEMENT FUSION CONCEPT

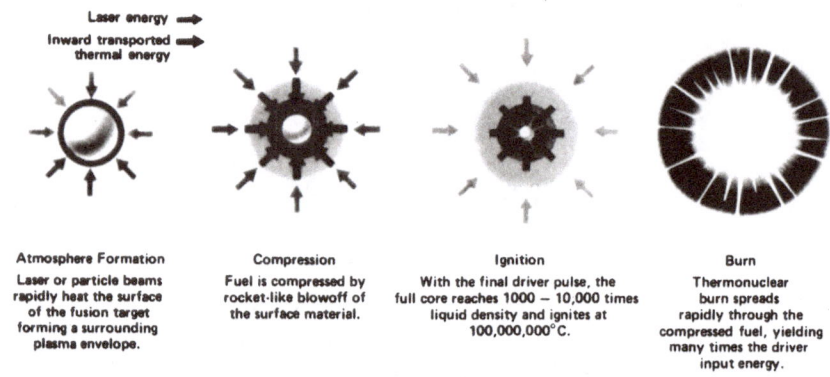

Atmosphere Formation	Compression	Ignition	Burn
Laser or particle beams rapidly heat the surface of the fusion target forming a surrounding plasma envelope.	Fuel is compressed by rocket-like blowoff of the surface material.	With the final driver pulse, the full core reaches 1000 — 10,000 times liquid density and ignites at 100,000,000°C.	Thermonuclear burn spreads rapidly through the compressed fuel, yielding many times the driver input energy.

Fig. 4.   Schematic diagram illustrating the principle of inertial confinement fusion. The diagrams illustrate the various stages of pellet compression, ignition, and burn.

3. Fusion Reactors

3.1   Reactor Design and Technology

Fusion reactor design studies began in the late 1960s, and the first specialist conference was held in 1969 at Culham.[2] By the early 1970s reference designs existed for many magnetic confinement systems and some intertial confinement schemes. Since then there has been a large increase in reactor studies with clear problem definition, detailed design work, and attempts to introduce more realism into the designs.

Basically, a D–T reactor comprises a plasma containment vessel, from which most of the energy from the reactions escapes in the form of fast neutrons to be captured in a blanket incorporating a cooling system, with heat exchangers leading to a conventional steam cycle. The blanket contains lithium so that tritium, which is not naturally occurring, can be regenerated. In a magnetic fusion reactor. which in principle can operate

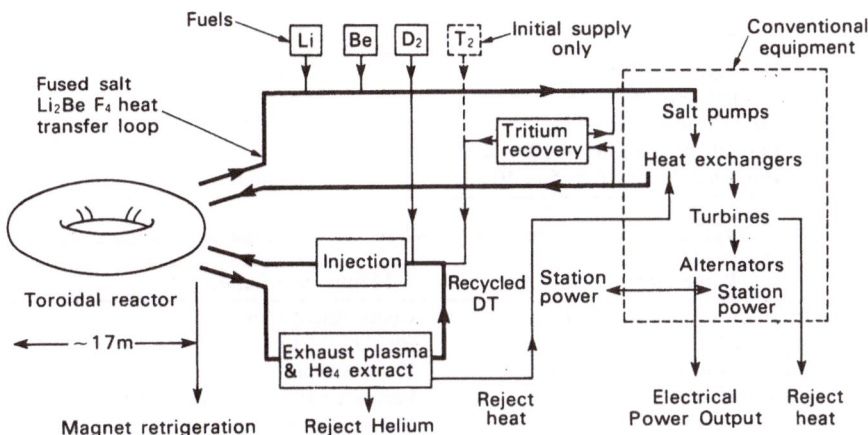

(a) a toroidal magnetic fusion power plant.

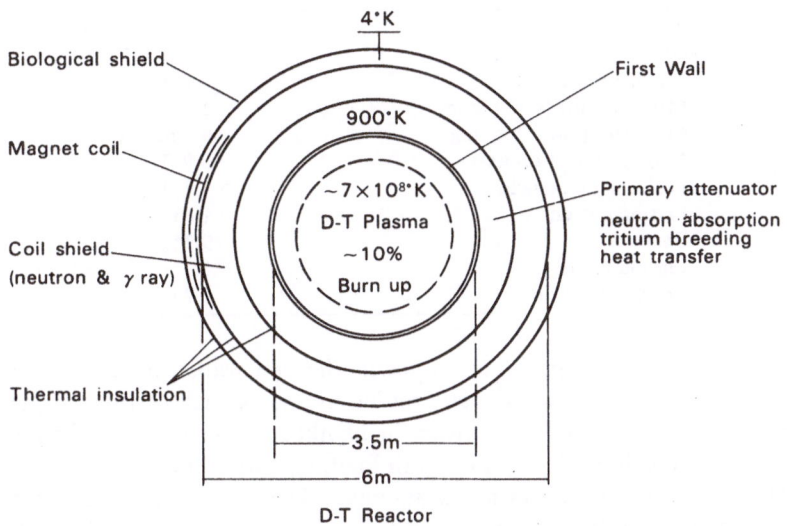

(b) the cross-section of a toroidal magnetic fusion reactor.

Fig. 5.   Schematic diagrams.

in steady state, the magnetic field coils, usually superconducting, are positioned outside the blanket protected by shielding. A schematic diagram of a toroidal magnetic fusion power plant and a cross-section of the reactor are shown in Figures 5(a) and 5(b).

Table 1. Typical plasma parameters of toroidal D-T reactor using magnetic confinement.

Parameter	Mainly determined by
Density, $n \sim 10^{14}$ cm^{-3}	Wall loading
Temperature, $T \sim 10 - 20$ KeV	Nuclear cross-section
Confinement time $\sim 1 - 10$s	Lawson criteria
Lawson product, $n\tau \sim 10^{14} - 10^{15}$ cm^{-3}s	
Efficiency factor, $\beta > 5 - 10\%$	Economics

Table 2. Typical machine parameters of D-T reactor using magnetic confinement.

Parameter	Mainly determined by
Minor radius, $a \sim 1 - 3$ m	Confinement time
Plasma radius $\gtrsim 1 - 3$ m	Blanket thickness
Major radius, $R \sim 5 - 10$ m	Power output
Magnetic fields $\sim 50 - 100$ kG	Pressure balance, strength of materials
Toroidal current $\sim 10 - 25$ MA	Heating, pressure balance
(pinch or tokamak)	

Table 3. Parameters of Culham MKIIB tokamak reactor.

Net electrical power	1,200 MW(e)
Gross thermal power	3,820 MW(th)
Major radius	6.7 m
Minor radius (mid-plane)	1.95 m
First wall power loading	4.7 MW/m^2
Magnetic field on axis	3.9 T
Peak magnetic field	7.7 T
Plasma current	10.3 MA
Beta	9.3%
Plasma density	3.2×10^{20} m^{-3}
Confinement time	1.7 s

One of the major problems concerns the first wall, which is subject to intense fluxes of electromagnetic radiation and particles from the plasma and back-scattered radiation from the blanket. The economics of a reactor are sensitively dependent on the wall loading, and in most designs a figure between 1 and 4MW per m^2 is assumed. The main plasma and reactor parameters can be determined relatively simply without the detailed knowledge of plasma physics and are shown for a toroidal D-T reactor in Tables 1 and 2. As an example of a tokamak reactor study, the parameters of the Culham MKIIB design[3] are given in Table 3. The net electrical power in this case is 1,200MW(e) and in general the output of a toroidal reactor is rather large. Figure 6 shows a schematic diagram of a new U.S. design, known as Starfire,[4] in which an attempt has been made to simplify the engineering and problems such as remote handling and

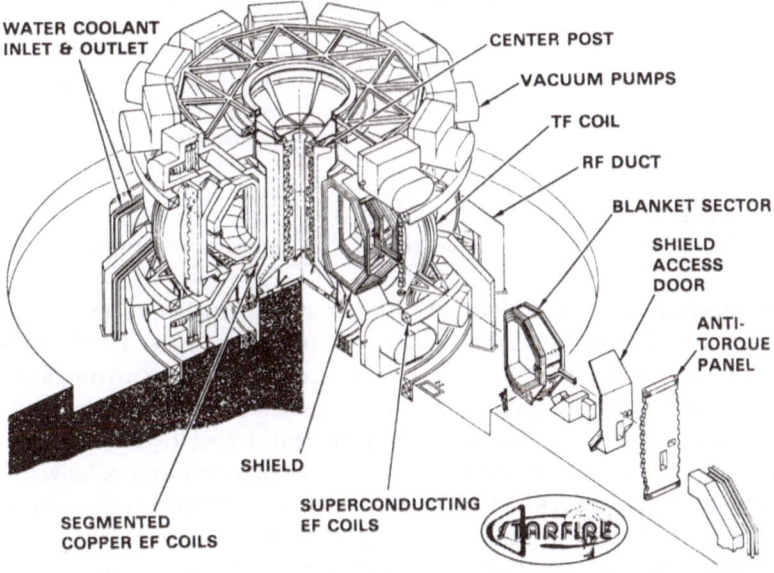

WATER COOLANT
INLET & OUTLET

CENTER POST

VACUUM PUMPS

TF COIL

RF DUCT

BLANKET SECTOR

SHIELD
ACCESS
DOOR

ANTI-
TORQUE
PANEL

SHIELD

SEGMENTED
COPPER EF COILS

SUPERCONDUCTING
EF COILS

Fig. 6.   Isometric view of the Starfire magnetic fusion reactor (Argonne National Laboratory, U.S.A.).

maintenance have been considered. Corresponding studies have taken place in Japan.

Extensive studies have been reported[5] on linear magnetic mirror reactors. These have considerable advantages because they are less complex with easy access and automatic removal of reaction products, but due to end losses they are only marginally economic unless operated as a hybrid (see Section 3.2).

Many attempts to estimate the costs of a magnetic fusion reactor and make comparisons with fission systems have been made. This is difficult because fusion is still in a relatively early stage of development and the results are sensitive to the wall loading and plasma β assumed. Because the power density in the "nuclear island" is relatively low for fusion, this part costs much more—by a factor ~ 8 according to a recent U.K. study—than its fission counterpart. Taking into account that most of the conventional plant is similar, a fusion power station is estimated to cost typically 2—4 times more than a PWR fission station. It is even more difficult to estimate how this ratio will vary in the future, since it depends on many factors including the world energy demand and uranium prices.

The main parameters of a laser-driven fusion reactor are given in Table

Table 4.   Typical parameters of D-T laser pellet reactor.

Parameter	Mainly determined by
Cavity radius $\sim 1 - 2$ m	Shock and thermal stress
	From micro-explosion
Laser energy $\sim 10^5 - 10^6$ J	Practical upper limit
Pellet mass $\sim 10^{-3}$ gm	Laser energy
Pellet initial diameter ~ 0.5 mm	Laser energy
Fusion energy yield/pellet $\sim 10 - 100$ MJ	Pellet design and size
Repetition rate $\sim 10 - 100$ per second	Mean power output

4, as representative of an inertial fusion system. A pellet of D-T initially of diameter of about 0.5 mm is compressed and heated in $\sim 10^{-10}$ s by laser light or a charged particle beam driver in a suitable cavity surrounded by a neutron-absorbing blanket. Such micro-explosions must be repeated ten or more times per second, and each pellet must cost between 0.1 and 1 U.S. cent. Problems include containing the micro-explosion, developing drivers with high enough efficiencies at high power and energy, and protecting the laser and its optics, or the particle beam source, from damage by debris from the explosions, and finding a suitable pellet injector. Each micro-explosion yields about 100 MJ (equivalent to 15 kg of high explosive, although the blast damage is less because the energy released is largely in radiation and neutrons rather than kinetic energy of fragments). The inner layer of the cavity wall will experience large temperature rises, thermal shock, and fatigue effects from 4 MeV alpha particles and intense X-rays. Most designs utilize a liquid lithium first wall, possibly in the form of a vortex. A schematic diagram of a recent Livermore[6] design of a laser fusion reactor is given in Figure 7. A new reactor scheme with magnetically confined flowing lithium was recently proposed in Osaka.[7]

Finally, we mention some of the technological problems in fusion reactor design, summarized in Table 5. Effects induced by 14MeV neutrons are much severer than those from fast fission neutrons at $1 - 2$MeV. Extrapolations are uncertain, and sufficiently intense sources of 14MeV neutrons to provide the required total radiation dose for large samples are not yet available. The most likely sources are large fusion experiments or prototype reactors. Major research programs on the technological

Table 5.   Technological problems of fusion reactors.

1. First wall design
2. Tritium handling and blanket design
3. Magnet design and superconducting technology
4. Material problems, especially for 14MeV neutrons
5. Power supplies
6. Reactor maintenance

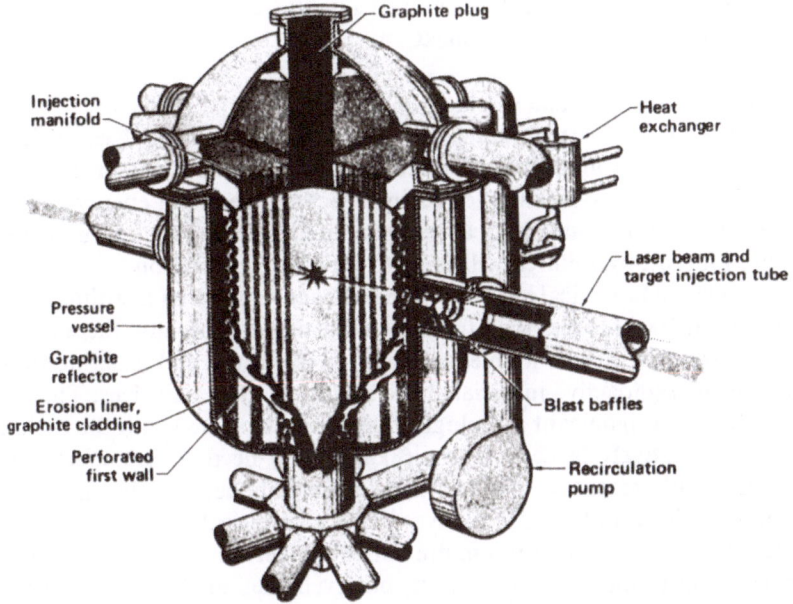

Fig. 7. The Hylife laser fusion reactor (Lawrence Livermore Laboratory, U.S.A.).

problems of fusion reactors are growing up in laboratories throughout the world.

3.2 Fusion-Fission Hybrid Systems

The idea of the fusion hybrid[8,9] is to surround a fusion reactor with a blanket of ^{238}U or ^{232}Th and utilize the fast fusion neutrons to breed fissile material, ^{239}Pu or ^{233}U, to fuel conventional reactors, possibly up to ten. In addition, each fission reaction produces about ten times the energy of a fusion reaction and releases several fission neutrons so that there can be a substantial energy gain and the system can breed. The requirements for T and $n\tau$ are less stringent than for a pure fusion reactor, and a hybrid could operate with a fusion component which does not by itself produce useful economic power. Hybrids could, therefore, be introduced before pure fusion is economic and are attractive for systems such as mirror machines where a pure fusion system is marginal. On the other hand, many of the environmental advantages of a pure fusion reactor would be lost.

One interesting scheme,[10] the so-called fission-suppressed hybrid, has improved environmental features and might fuel as many as 25 light water reactors. In this system the blanket composition is optimized for neutron multiplication and breeding using Be and ^{7}Li rather than for fast fission and energy multiplication.

3.3   Fusion Reactors and the Environment

Studies of fusion reactors and the environment have been carried out[11,12] and it is found that there are simpler solutions to the problems of safety and radioactivity than for fission reactors. The total energy content of the plasma is small and a self-generated nuclear explosion cannot occur; indeed any failures will tend to quench the reaction.

The relative biological hazard of the radioactive material is estimated [11] to be about 100-fold less than for fission, particularly because there are no fission products such as iodine, caesium, or strontium and no plutonium; most of the radioactive materials have relatively short half-lives. There is, nevertheless, a significant radiological problem associated with the tritium, of which relatively large amounts (3kg/GW(e)) need to be safeguarded and the whole reactor structure will become radioactive. The maximum credible accident (a highly unlikely event) is a large-scale release of tritium, probably as a result of a lithium fire in the blanket; this could release more than 10^4 curies into the atmosphere, but even such an accident would be relatively local in its effects.

4. Magnetic Confinement Research

4.1   Tokamaks

The tokamak (Figure 1) is derived from the historically important pinch effect[13] in which a plasma column is confined by the self-magnetic field, B_θ, of a current. The simple pinch was first studied in toroidal geometry by Cousins and Ware[14] in 1950 in the U.K. Its behavior is dominated by magnetohydrodynamic instabilities (Figure 3) which are reduced by the addition of a small toroidal magnetic field, B_ϕ. This so-called "stabilized pinch" was the most widely studied system, both in linear and toroidal geometry, in the early days of fusion research between 1955 and 1960, after which it was abandoned in most laboratories because stability could not be obtained. In the Soviet Union it was found that stable conditions could be realized by applying a large toroidal field such that $B_\phi \gg B_\theta$; more quantitatively, it was observed that stability could be obtained when:

$$q = aB_\phi/RB_\theta \geq 3$$

where R and a are the major and minor radii of the torus. This step was the beginning of the tokamak program.

By about 1965, temperatures of a few hundred eV, with confinement times of about l ms, were reported. By 1968, in the Kurchatov Institute in Moscow, using a device called T3 (R = 100cm, a = 15cm, $B_\phi \leqslant$ 34kG, I $\leqslant$ 120kA) supported by some smaller machines, electron temperatures (T_e) exceeding 1keV and confinement times of several milliseconds were obtained[15] with thermonuclear neutron emission. The foundation of modern tokamak research was laid using these machines. The temperature was first estimated indirectly from electrical measurements. However, in 1969 a team from Culham Laboratory (U.K.) went to Moscow with newly developed equipment to measure temperature by the scattering of ruby laser light and confirmed that high values were reached.[16] Following this, tokamaks were built in many fusion laboratories throughout the world, including JET-2 and DIVA in Japan (see review articles, Ref.[19]).

In the period 1970–75 the results from the Russian experiments were confirmed and extended on many different machines. In the early part of this period, plasma was heated by ohmic dissipation from the current but this becomes inefficient at high temperatures because the plasma resistivity falls off as $T_e^{-3/2}$. In about 1975 additional heating was investigated, first by the injection of high-energy beams of neutral atoms and later using a variety of high-frequency waves, for example at the cyclotron frequency of the ions or the electrons. In this period the basic results from T3 and its successor T4 were confirmed, and many new problems studied including investigations of compressional heating, the use of a non-circular cross-section which can lead to higher values of β, studies using very high magnetic fields which permitted high values of the current density and more effective ohmic heating, density control using pulsed gas valves, and new techniques for cleaning the walls of the toroidal vessel to reduce high atomic number impurities which give large energy losses by radiation. The first experiments using diverters were reported, from DITE at Culham[20] and from DIVA[18] at JAERI. This is a device which enables the outer layers of the plasma, contaminated with impurities from the walls, to be scraped off and removed to a secondary chamber and pumped away. By 1975 many of the basic properties of the tokamak were established with plasma parameters: $T \sim 2 - 3$keV, $n \gtrsim 10^{14}$cm^{-3}, confinement times up to about 10ms, with $n\tau \sim 10^{12}$cm^{-3}s; the value of β, however, remained less than 1%.

The most important question concerns the confinement of the plasma, which can lose energy to the walls by cross-field ion or electron heat conduction. In general, the ion contribution agreed with so-called neo-classical collisional theory, but that from the electrons, which should be smaller by the square root of the mass ratio, was anomalously large by a factor of 10–

100. There is still no adequate explanation for this discrepancy, which is less at high density. An important scaling law for the confinement time, τ, was established on Alcator[21] at MIT (U.S.A.) as follows:

$$\tau \propto (\text{plasma density}) \times (\text{minor radius})^2.$$

Data from most tokamaks fit this empirical relationship within a factor of 2 or 3, as seen in Figure 8, but there is no satisfactory theoretical explanation for it. This dependence is quite different from the theoretical one given in Section 2.2.

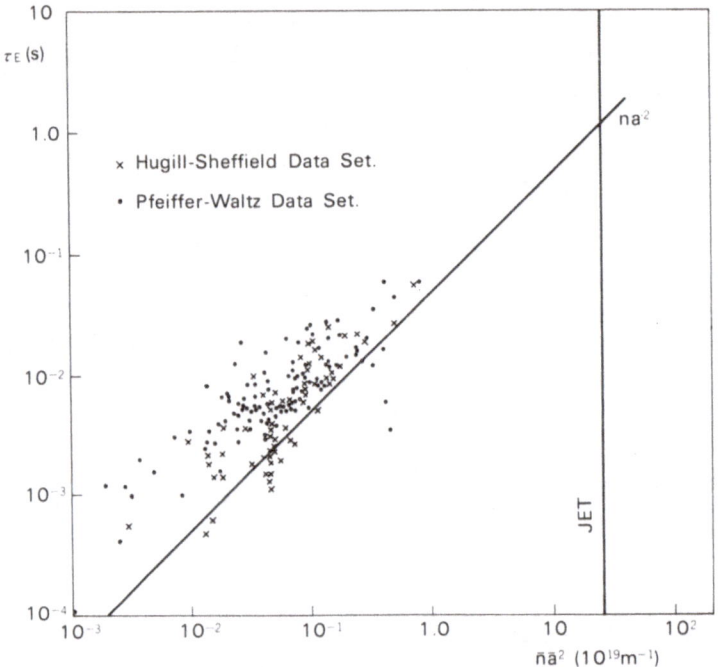

Fig. 8.   Empirical tokamak scaling—the confinement time is shown as a function of $\bar{n}\bar{a}^2$ (n is the particle density and a the minor radius of the torus).

The period 1975–80 saw several major advances, including the operation of two much larger tokamaks, PLT[22] at Princeton (R = 1.3m, a = 0.45m, Bϕ = 50—60kG, I $\sim$ 600kA) and T10[23] at the Kurchatov Institute (R = 1.5m, a = 0.39m, Bϕ = 45kG, I $\sim$ 650kA). Figure 9 shows a photograph of PLT. These two machines gave much important information, in particular to confirm the dependence $\tau \propto a^2$, on a larger

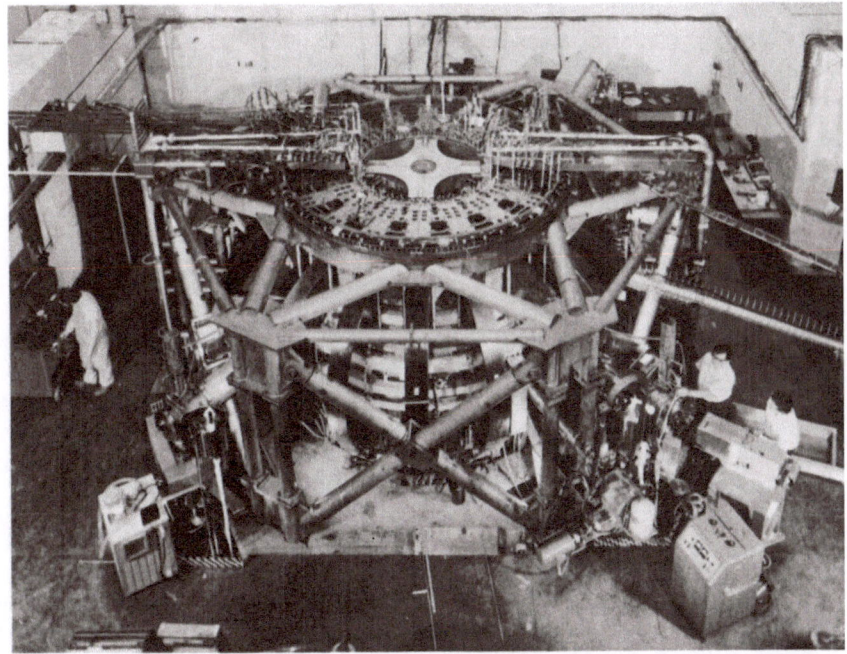

Fig. 9.   A photograph of the Princeton Large Torus (PLT).

system with $\tau \sim 50$–100ms; the plasma lifetime, during which the energy is replenished many times by ohmic or additional heating, extended up to $\gtrsim 1$s. A very large non-circular cross-section tokamak, Doublet III,[24] shown schematically in Figure 10, began operation at the General Atomic Laboratory in San Diego in 1978, with experiments carried out by teams of American and Japanese scientists, as part of a collaborative programme between these two countries. There are two large new experiments with divertors, PDX[25] at Princeton and ASDEX[26] in Garching, from which the first results were reported this year.

We conclude the account of tokamak research by giving examples of the most outstanding results which have been reported recently. The highest temperature, reached on PLT,[22] is 7.1keV, which exceeds the minimum ignition temperature. This was obtained using $\gtrsim 3$MW neutral injection at 40kV and a H^+ ion spectrum is shown in Figure 11 (a). In Figure 11 (b), the temperature-time variation is shown. The highest average value of

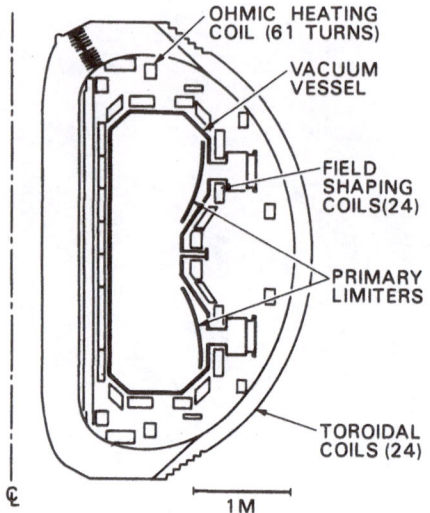

Fig. 10. A cross-sectional view of the Doublet III machine at General Atomic Laboratory, San Diego.

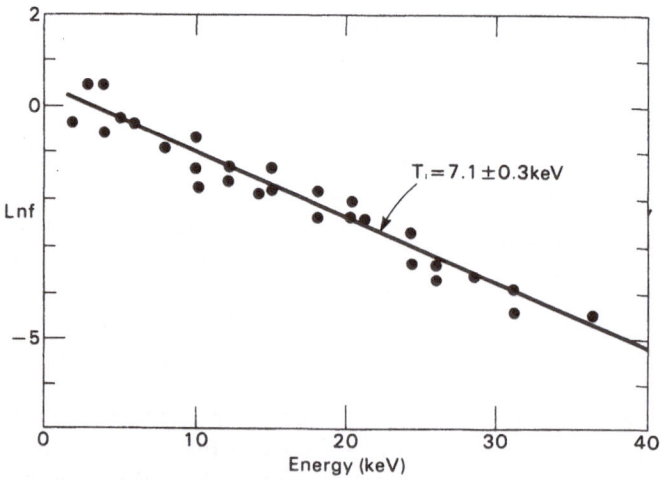

Fig. 11. (a) Hydrogen ion spectrum obtained on PLT with neutral injection.

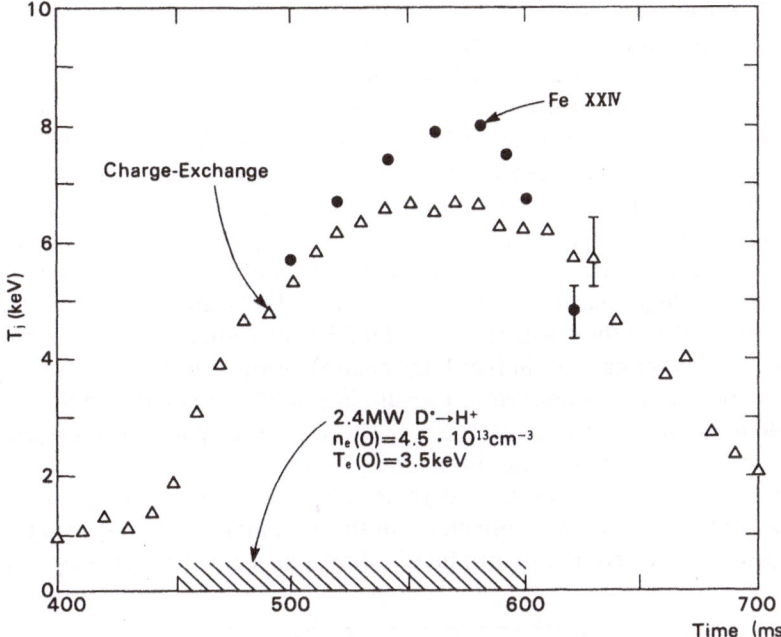

Fig. 11. (b) Ion temperature versus time on PLT with neutral injection. The temperature was determined by charge-exchange measurements.

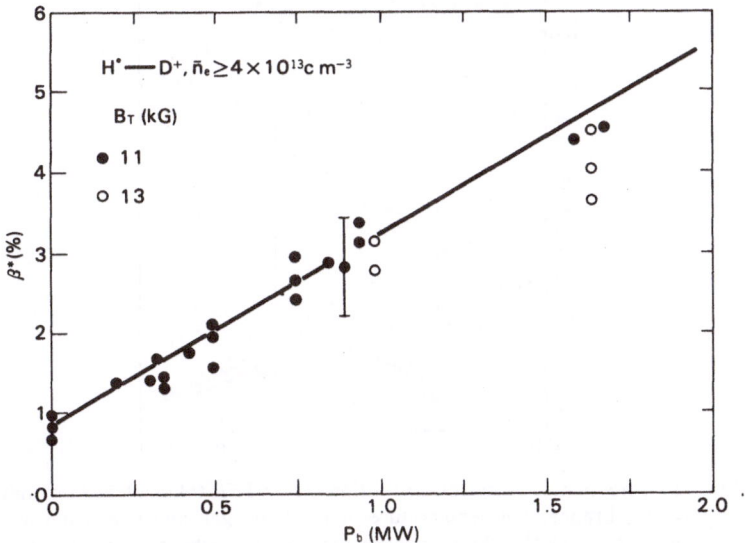

Fig. 12. Average value of the efficiency factor, β^*, as a function of neutral injected power obtained on the ISXB tokamak at Oak Ridge.

β obtained, on a smaller machine called ISXB at Oak Ridge,[27] also using neutral injection, is $\backsim 4\%$; data are shown in Figure 12. Similar values of β have been obtained on JFT-2[28] at JAERI.

High-frequency heating, with power levels up to ~ 1MW has produced effective heating, as evidenced by the data from TFR[29] in France, shown in Figure 13. The highest temperatures reached with wave heating is $\gtrsim 2$keV on PLT.[22] Important results have also been obtained using different kinds of wave heating on DIVA[28] and JFT-2[30] at JAERI. The highest value of n$\tau \sim 3 \times 10^{13}$cm^{-3}s, has been obtained on the Alcator[21] high-field, high-density tokamak with $T \gtrsim 1$keV and peak n $\backsim 1.5 \times 10^{15}$cm^{-3}. Recently it was shown in DITE[31] at Culham (Figure 14) that a plasma current can be induced by neutral beam injection, as predicted theoretically; this is important because in principle it could lead to steady-state tokamak operation. Methods of the refueling system by the injection of small pellets have also been developed.[27]

In summary, over the last 20 years there has been steady progress in increasing the plasma parameters on the tokamak. This is illustrated in Figure 15(a) where the nτ product is shown as a function of temperature

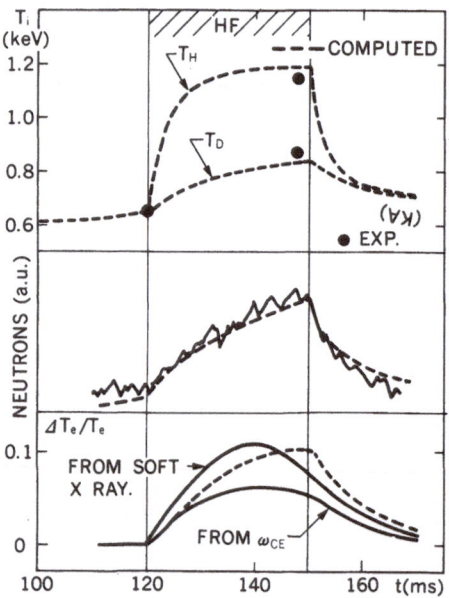

Fig. 13.   High-frequency ion cyclotron heating on the TFR tokamak at Fontenay-aux-Roses, France. The temperature of the hydrogen and deuterium ions, the electrons, and the thermonuclear neutron flux—theory and experiment, is shown.

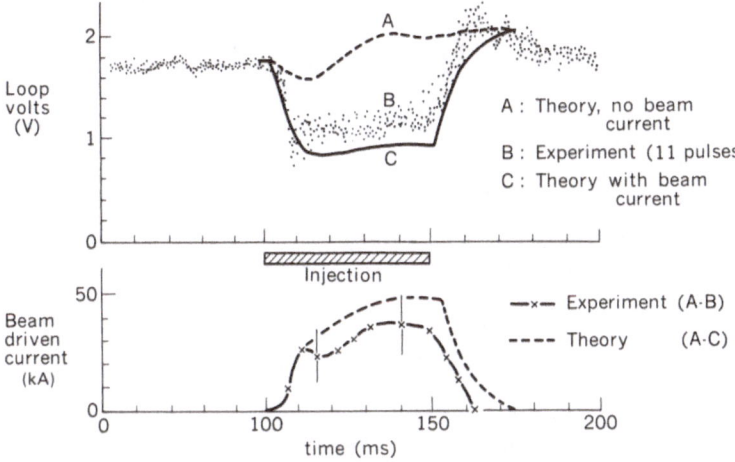

Fig. 14. Evidence for a uni-directional current driven by neutral beam injection obtained on the DITE tokamak at Culham: theoretical and experimental results are shown. The upper figure shows the loop voltage (which drives the plasma current) as a function of time. During injection the voltage is reduced due to the beam. The lower figure shows the current.

between 1955 and 1980, and in Figure 15(b) where the plasma lifetime and the ion temperature are shown as a function of calendar years.

4.2   Alternative Toroidal Systems

4.2.1   Reversed Field Pinch (RFP)

The reversed field pinch[32] (Figure 1) had its origins in the simple pinch[13] as did the tokamak. Compared with the tokamak, it can operate at higher values of the efficiency factor, β (Section 2.2), with lower toroidal fields and with larger currents so that in principle ignition by ohmic heating from the current without additional heating might be possible. These features might lead to a less complex reactor. The RFP possesses the interesting physical property that the plasma itself forms the desired stable field configuration by means of a self-stabilization process which occurs naturally, involving a redistribution of the plasma currents. This is explained because the configuration is a near-minimum energy state to which the plasma relaxes.

In 1968, on the large Zeta RFP experiment[33] in the U.K., the first example of prolonged toroidal confinement of a high-β plasma was reported. Relatively stable "quiescent" conditions were found with plasma

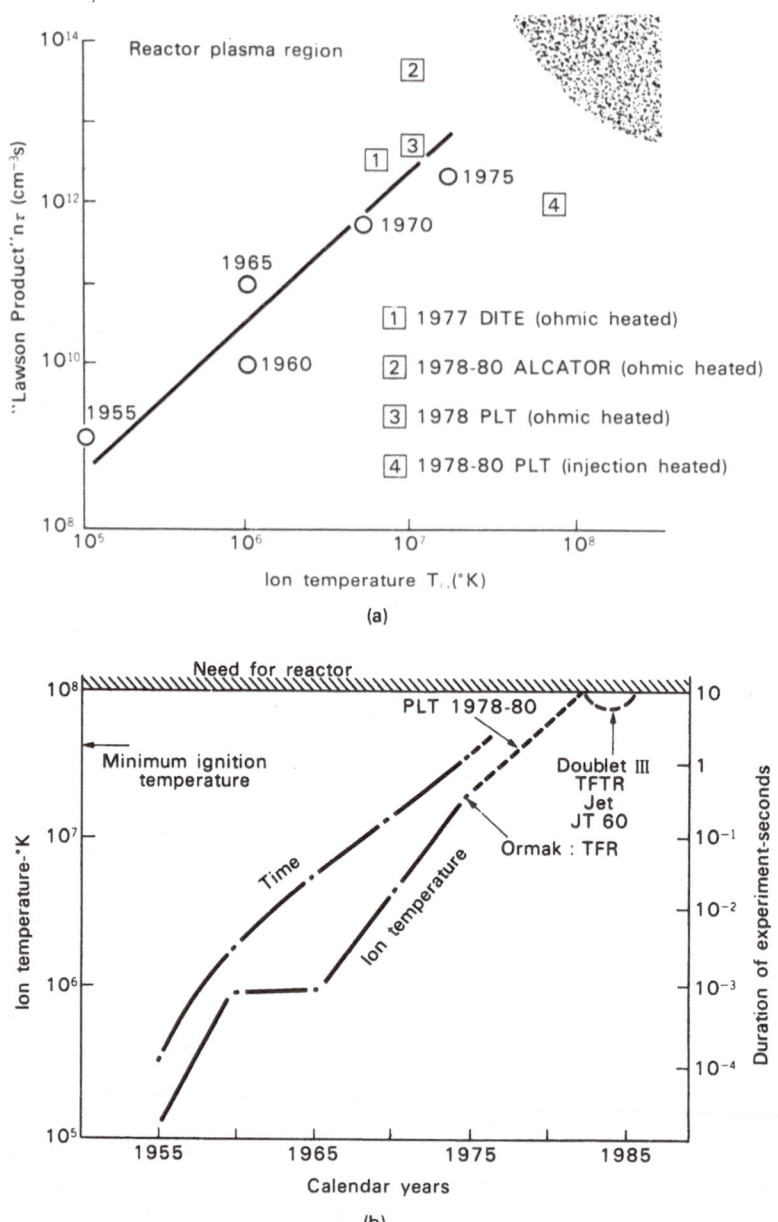

Fig. 15. (a) The Lawson product, $n\tau$, and ion temperature, T, obtained from tokamaks between 1955–1980.

(b) The ion temperature and duration of the experiment as a function of
calendar years from 1955–1980.

parameters: $T \sim 150 - 200$eV, $n \sim 5 \times 10^{13}$cm^{-3}, $\beta \gtrsim 10\%$, and a confinement time $\sim 3 - 10$ ms (comparable to tokamaks of that era). In the 1970s RFP research continued with smaller apparatus in which the field configuration could be set up by fast magnetic field programming and many theoretical predictions were confirmed experimentally at low temperatures. In these small experiments the quiescent conditions observed in Zeta were not found. Recently in Eta Beta II[34] at Padua (Italy) and in TPE-1R(M)[35] at the ETL Laboratory in Japan, quiescent behavior similar to that in Zeta was reported with improved confinement and effective ohmic heating to $T \sim 100$eV.

Reversed field pinches are now being studied at Culham, Padua, Los Alamos in the U.S.A., and at Nagoya and the ETL Laboratory. The two new intermediate-sized devices called HBTXIA (Culham) and ZT-40 (Los Alamos) which are larger than the Padua and ETL machines, will shortly be operating (ZT-40 has already operated but in a different regime). It is also planned to build at Culham a much larger RFP called RFX, in collaboration with Padua and Los Alamos, whose objective is to find out if temperatures up to 1keV can be obtained in a self-stabilized RFP discharge at high-β by ohmic heating. Calculations indicate that in order to obtain such temperatures a current ~ 2MA is required. The RFX design parameters are: current $= 2$MA, major radius $= 180$cm, minor radius $= 60$cm, timescale $= 0.25$s. An artist's impression of the machine is shown in Figure 16. The design is complete and international approval is being sought; the machine will take about 5 years to build.

4.2.2 Stellarators

The stellarator (Figure 2) was proposed at Princeton[36] and was one of the first systems to be considered for a reactor. Its main advantage is that it can be steady state but it is more complicated than a tokamak. In the 1960s there were large programs in stellarator research at Princeton, Garching (FRG), in the U.S.S.R., and in Japan. However, partly because of the unsuccessful results obtained from the large C-stellarator at Princeton, for reasons which have never been fully explained, stellarators were abandoned in the U.S.A. Work continued at Garching, at the Lebedev Institute and at Kharkov, at Culham, and at Nagoya and Kyoto.

In 1975 a major advance in stellarator research was reported from Culham,[37] Garching,[38] the Lebedev Institute,[39] and Nagoya,[40] when it was found that by passing a current round the stellarator and operating in a regime which was part stellarator and part tokamak, effective ohmic heating and good confinement similar to that observed in tokamaks could be obtained. In the optimum case, plasmas with $n \gtrsim 10^{13}$cm^{-3}, $T \gtrsim 500$eV and confinement time $\gtrsim 5$ms, could be sustained for times $\gtrsim 100$ms. There were indications that the confinement improved when the current

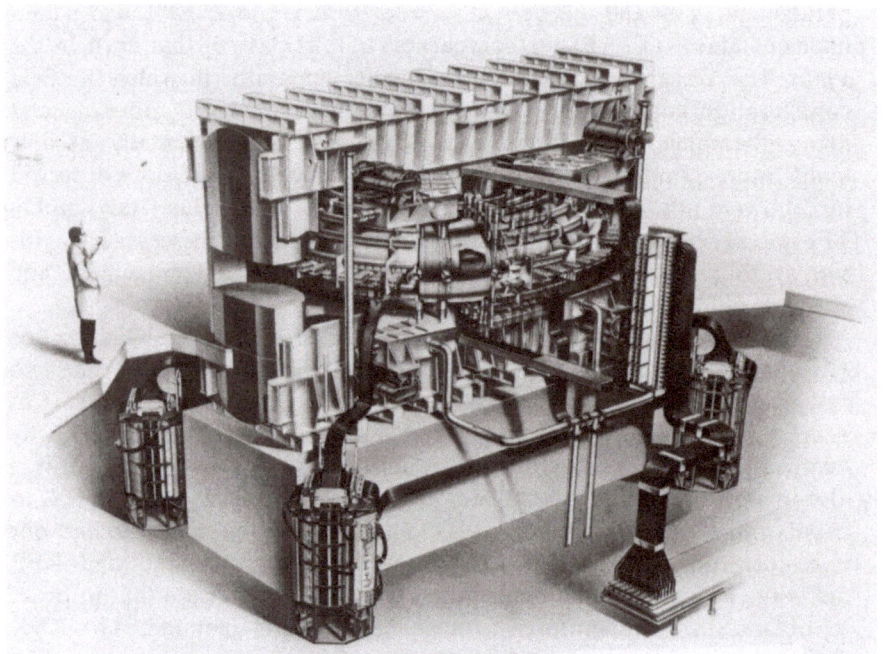

Fig. 16. Artist's impression of the proposed RFX reversed field pinch experiment.

became smaller. A second advance was made on the Wendelstein VIIA stellarator at Garching[41] in 1980 using 100–200 KW of neutral beam heating. A well-confined plasma with $T \sim 300$–500eV could be maintained by the beam while the current was reduced approximately to zero. This demonstrated the original concept of a currentless stellarator. In Japan neutral injection and high-frequency heating studies are under way in JIPP T-II.[42]

The future outlook for stellarator research is uncertain. Although there is now a renewed interest in the U.S.A., in Europe the Culham stellarator program has been closed down, but at Garching there are plans for a successor to Wendelstein VIIA. The Soviet Union is continuing stellarator research as also is Japan where there is the largest stellarator yet to be constructed, the Heliotron E at Kyoto.[43]

4.2.3 Other Toroidal Systems

There are many other toroidal systems such as the compact tori (including screw pinches and high-β tokamaks), and the bumpy torus, studied in the U.S.A.[44] and in Japan.[44] There are several interesting new ideas yet to be fully developed, such as the spheromak[45] at Princeton, and the OHTE concept, which is a kind of RFP-stellarator hybrid, proposed at

General Atomic,[46] and the use of a relativistic electron ring to provide confinement, studied in Nagoya.[47]

4.3   Linear Systems

4.3.1   Magnetic Mirrors

In the simple mirror the main confining field is produced by a uniform solenoid with the field at the ends increased so that gyrating particles are reflected, and this reduces end losses. The mirror principle operates in the Van Allen belt. In practice a more complicated scheme called a minimum-B mirror or a magnetic well is used. Calculations show that because of end

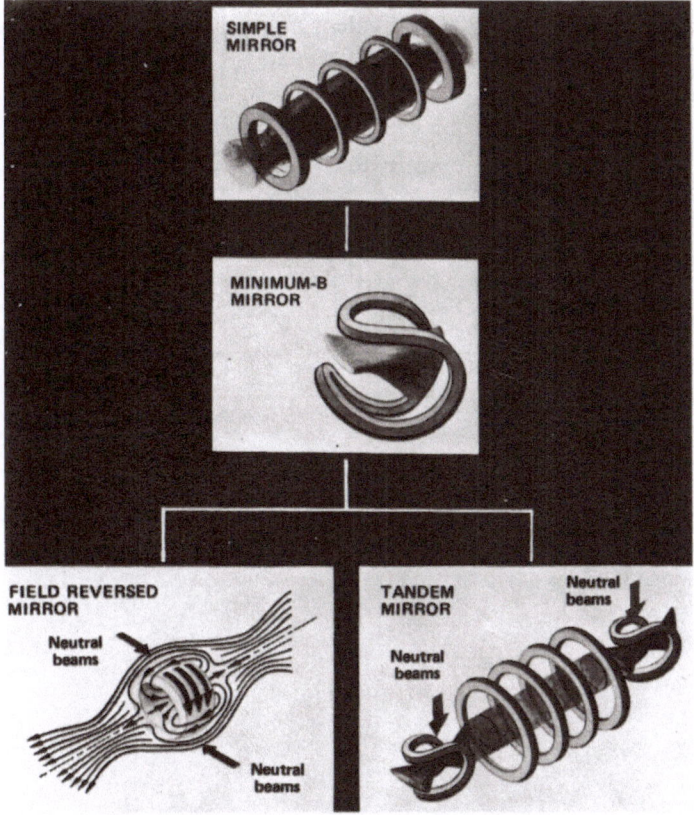

Fig. 17.   Schematic diagrams of variants of the magnetic mirror machine. The coil arrangement is shown for the simple mirror, the minimum-B mirror, and the tandem mirror using electrostatic end plugs and the field configuration is shown for the reversed field mirror (Lawrence Livermore Laboratory).

losses the mirror system is at best marginal for an economic reactor and at Livermore (U.S.A.) there have been many ingenious ideas for reducing the end losses, including the field reversed mirror in which a closed-line configuration is generated within the system, and the tandem mirror in which electrostatic potential barriers are developed at the ends which give improved confinement. These mirror variants are illustrated in Figure 17.

The mirror machine[48] was proposed at the Lawrence Livermore Laboratory (U.S.A.) where it has been studied since fusion research began; a large program also developed in the Soviet Union. In most other countries mirrors were abandoned some time ago, because of end losses, in favor of toroidal systems. At Livermore there is a large tandem mirror, called TMX,[49] in operation (see Figure 18), and an even bigger device, called MFTF, is under construction. In TMX there are 24 neutral beams with currents exceeding 200 equivalent amps at about 15kV at each end

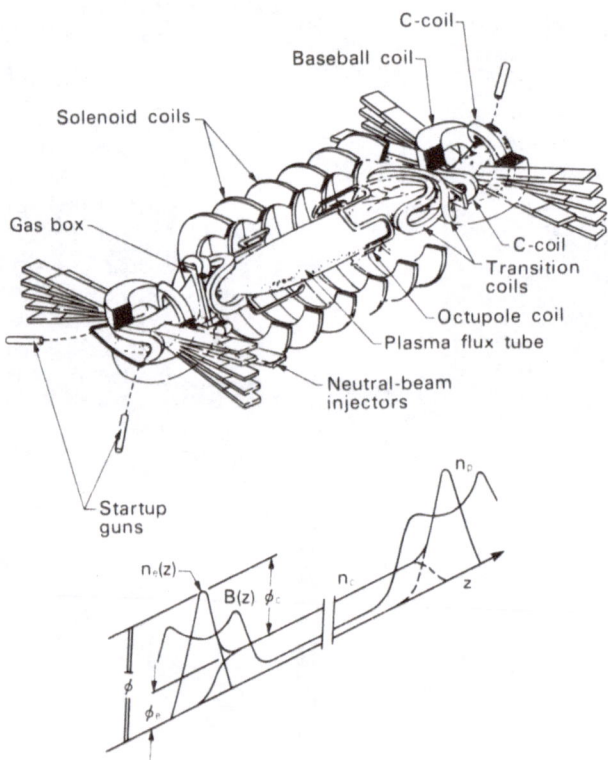

Fig. 18. Schematic diagram of the TMX tandem mirror machine at Lawrence Livermore Laboratory showing also the neutral beam system. The lower diagram shows the magnetic field, density, and electrostatic potential profiles.

cell. The plasma parameters are considerably higher than in previous experiments and are, in the central cell, electron temperature $\sim 260eV$, density $\sim 3 \times 10^{13}cm^{-3}$, $\beta \sim 0.1$—0.5, and confinement time a few milliseconds. The ion energy in the end cells is $\sim 15keV$. The effectiveness of the electrostatic plugs is illustrated in Figure 19, which shows that if one of them is turned off the losses through that end rise sharply. It may be concluded that there have been promising advances in mirror machine research in the U.S., but there is still some way to go before tokamak-like parameters are achieved.

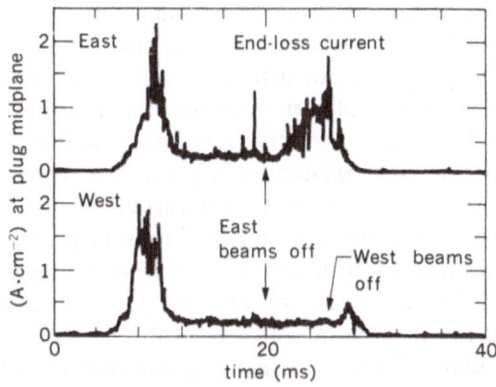

Fig. 19.   The effect of the electrostatic end plugs in the TMX mirror machine at Livermore. When the east plug is turned off at 20ms, the current flowing out of the east end increases relative to that from the west end indicating the effect of the west end electrostatic plug.

4.3.2   Theta Pinches and Other Linear Systems

No account of fusion research would be complete without mentioning the theta pinch, in which a rapidly rising magnetic field ($\sim 50kG$ in a few μs) generated by a single-turn solenoid is used to heat a plasma by fast magnetic compression. Near thermonuclear conditions, $T \sim 5keV$, $n \sim 10^{16}cm^{-3}$, $\beta = 0.5$–1, were readily produced for a few microseconds. However, the theta pinch was limited by end losses, and attempts to stopper the ends by material plugs or to find stable toroidal theta pinch equilibria have met with only limited success. A few variants of the theta pinch are still studied, and there is some research on the related laser-heated solenoid in which a linear plasma column is heated by a powerful laser beam. One scheme which has received special attention in Japan is the use of radio frequency end plugs[50] to reduce the end losses in linear systems of the magnetic mirror type.

5. Inertial Confinement

The idea of ablative compression driven by laser beams (Section 2.3) to achieve a thermonuclear reaction in superdense matter was first reported in 1972.[51] Electron and ion beam driven fusion was considered later. Most of the experiments carried out so far and all those at very high power levels have used lasers, usually neodymium glass at $1.06\mu m$ or carbon dioxide at $10.6\mu m$. In the late 1970s pellet compression and neutron emission using electron beams was demonstrated but most of the work with particle beams is still on the development of large drivers.

The pellet compression process is very complicated. The main problem is the controlled transport of energy into the core to give both the required heating and compression together. Many interesting physics problems have been studied, such as hydrodynamics and shock phenomena, implosion dynamics and Rayleigh Taylor instabilities, the effects of non-uniform irradiation, the generation of fast electrons which will heat the pellet core and prevent high compression, the generation of fast ions which escape and remove energy, and magnetic field generation which may inhibit heat transport. Laser light absorption has been studied extensively and involves classical inverse bremsstrahlung and excitation of parametric instabilities, in particular stimulated Brillouin scattering at the critical density (n_c)* and stimulated Raman scattering at $n_c/4$. These processes play an important role in the physics of coupling beams to the plasma. Non-linear coupling from density profile steepending can lead to fast particle generation.

In addition to pellet compression studies, there has been extensive work developing different types of diagnostics, including X-ray photography, spectroscopy, neutron diagnostics, and radio-chemical techniques.

The first results on pellet compression were obtained from an elegant series of experiments carried out at KMS Fusion in the U.S.A. in 1973–74.[52] A simple optical arrangement was used so that approximately uniform radiation of the pellet was possible using only two laser beams. The pellets comprised small thin-walled glass micro-balloons of typical diameter 50—100 microns filled with D-T gas with a pressure $\sim$ 15 atmospheres. The laser energy was a few hundred joules. About 50J fell on the target corresponding to a power of about $10^{15}W\ cm^{-2}$ with a pulse width $\sim$ 3×10^{-10}s. About 10J were absorbed. Temperatures approaching 1keV, with neutron yields $\sim 10^5 - 10^6$ were obtained, increased later using more powerful lasers. Compressions of between 100 and 1,000 were deduced from X-ray photographs and comparison with numerical simulations.

---

* *The critical electron density occurs when the electron plasma frequency ($\omega_p \propto n_e^{1/2}$) equals the laser light frequency.*

These experiments demonstrated many of the ideas of pellet compression for the first time, and there was reasonable agreement with numerical codes. High compression ratios were produced but the targets used, called "explosive pusher" (see Figure 22) are not suitable for reaching the high absolute densities needed and the principle of ablative compression to ultra-high density, which also requires more powerful lasers, could not be demonstrated. Essentially with the targets used the gas was heated and compressed by a shock wave from the imploding shell. It was also found that less power went into the compression than had been expected, possibly because of fast ion production, the energy falling on the target was less than expected, and radiation losses from the glass may have played a role.

Between 1974 and 1980 laser fusion studies were carried out at many laboratories throughout the world, still usually using explosive pusher targets, and much more powerful lasers were built and used for pellet compression. The biggest program is at Livermore[53] in the U.S.A., using neodymium glass lasers and there is a large effort at Los Alamos[54] using CO_2 lasers, with work in other U.S. centres also. Work has been reported from the Lebedev Institute in Moscow using a 27-beam neodymium system. Studies in Europe,[55] in particular in France,[56] have not so far used lasers with as high powers as those at Livermore. There is a large programme at Osaka.[57] Particle beam-driven fusion is studied at the Kurchatov Institute[58] and in Osaka[59]; in the U.S.A. there are programs using electron and light ion beams at the Sandia Laboratory,[60] light ion beams at the Naval Research Laboratory and at Cornell University,[61] and feasibility studies on heavy ion beams at the Argonne Laboratory.[62]

In Table 6 the potential drivers for inertial confinement fusion are listed. Lasers have the advantage that they are easy to focus and to get high power on the target, but the efficiency tends to be small, especially at high powers. Electron and in particular ion beams deposit their energy more readily on the target but the incident power density is less since they are more difficult to focus. Very large energies are relatively easy to obtain; facilities in the megajoule range are under construction (one exists), and the efficiency is high. The required current for heavy ion beams is $\sim$ 10kA compared with 20MA for light ions and 100MA for electrons. Particle beam drivers do not require large extensions of existing technology. Although

Table 6.   Potential drivers for inertial confinement.

Lasers	CO_2(10.6 μm) HF (2.7 μm)
	KrF (0.24 $-$ 0.27 μm)
Heavy ions	20GeV, 10kA^{238} U$^+$
Light ions	10MeV, 20MA^2 H$^+$
Electrons	2MeV, 100MA

neodymium has been the most successful laser for experiments because very high powers are available, the efficiency is too small to be a potential reactor driver. The most likely laser systems for use in a reactor are CO_2 (10.6μm), hydrogen-fluoride (2.7μm) or krypton-fluoride (0.24—0.27μm). Efficiencies of 3—5% at high power are already obtainable and values of 5—10% possible. The krypton-fluoride in particular appears attractive be-

Table 7. Large laser fusion facilities.

U.S.A.						
	Operating on target		Under construction		Design work funded	
LLL	Shiva	10kJ	Nova I	10kJ	Nova II	300kJ
Nd/glass 1.06μm		26TW		150TW		300TW
LASL	Helios	5–10kJ	Antares	100kJ		
CO_2 10.6 μm		10–20TW		100–200TW		

Energy quoted reflects maximum performance at long pulses.
Power quoted reflects maximum performance at short pulses.

Japan					
		Number of beams	Power	Energy	Date
Osaka	GEKKO XII Nd/glass	12	40TW (0.1ns)	20kJ (1ns)	1982
Osaka	LEKKO VIII CO_2	8	19TW (1ns)	10kJ (1ns)	1981

Table 8. (a) Large electron beam fusion facilities.

Location	Voltage	Current	Pulse Half-width	Energy	Power
Angara V* (Kurchatov, U.S.S.R.)	2MeV	40MA	90ns	$\sim$ 5 MJ	80 TW
Aurora (U.S.A.)	1MeV	1.2MA	120ns	2.5MJ	20 TW
Reiden IV REB (Osaka, 1980)	$\sim$ 1.5MeV	$\sim$ 1.5MA	50ns	100kJ	2 TW

* One of 48 modules built.

Table 8. (b) Light ion beam fusion facilities.

Location	Operating on target	Under construction	Design work funded
Sandia	PROTO I–12kJ 0.5TW	PROTO II–160kJ 4.8TW PBFA I – 1MJ 30TW	PBFA II–2–4MJ 60–100TW
Osaka	REIDEN IV LIB–50kJ 1TW(50ns)		

Fig. 20.   Photograph of part of the Shiva laser system at the Lawrence Livermore Laboratory.

cause of its shorter wave-length. Iodine lasers have been studied in Garching, FRG.

Examples of large driver facilities both operational and planned, including those in Japan, are given in Table 7 (lasers) and Table 8 (particle beams). It is seen that both neodymium glass and CO_2 lasers with energies up to 10kJ have been used for pellet compression, and a 12kJ light ion system is operational at Sandia. A photograph of part of the Shiva laser is shown in Figure 20. The power of neodymium glass systems at Livermore is shown as a function of calendar years in Figure 21, and it is seen that progress has been rapid. There has been considerable development work on target design and construction. Schematic diagrams of different kinds of targets for laser fusion are shown in Figure 22; quite complicated

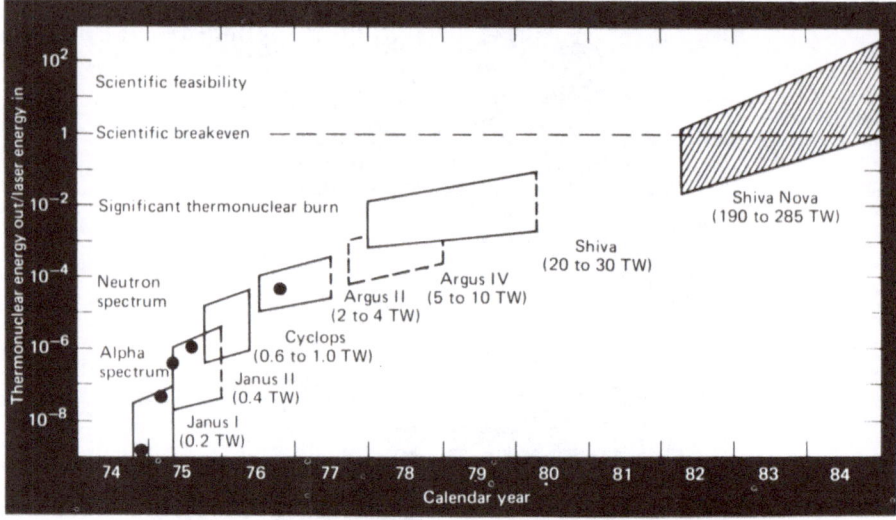

Fig. 21. Diagram illustrating the progress in laser fusion research at the Lawrence Livermore Laboratory. Neodymium glass lasers of increasing power are shown as a function of calendar years. The points illustrate experimental data up to 1977. Shiva has now given results also.

layered targets, which must be manufactured cheaply, are required to obtain high densities and a significant thermonuclear burn. The targets for particle beam drivers tend to be larger and differ somewhat in their detailed design.

The highest parameters reached using explosive pusher targets with the 10kJ Shiva laser has been $T \gtrsim 10\text{keV}$, neutron yield $\sim 3 \times 10^{10}$, pellet gain $\sim 1\%$, and $n\tau$ values $\sim 2 \times 10^{12}\text{cm}^{-3}$. Temperatures in the kilovolt range with intense thermonuclear reactions have also been observed using CO_2 lasers. Preliminary studies using Shiva with targets designed to give high densities have yielded values up to 100 times that of liquid D-T with $T \sim 0.5 - 1\text{keV}$. An X-ray photograph of an early high-density compression using the Helios CO_2 at Los Alamos is shown in Figure 23.

In summary, many features of the compression process are now in agreement with large numerical codes such as the Lasnex code used at Livermore. High temperatures have been produced at low density and high densities at lower temperatures, but not yet both together. It is now generally agreed that for a laser fusion reactor, a pellet gain of 100—1,000 is required, and this appears to be possible; laser energies in the range $10^5 - 10^6\text{J}$ are probably needed. The design and development work for such systems is

FUSION TARGET DESIGNS

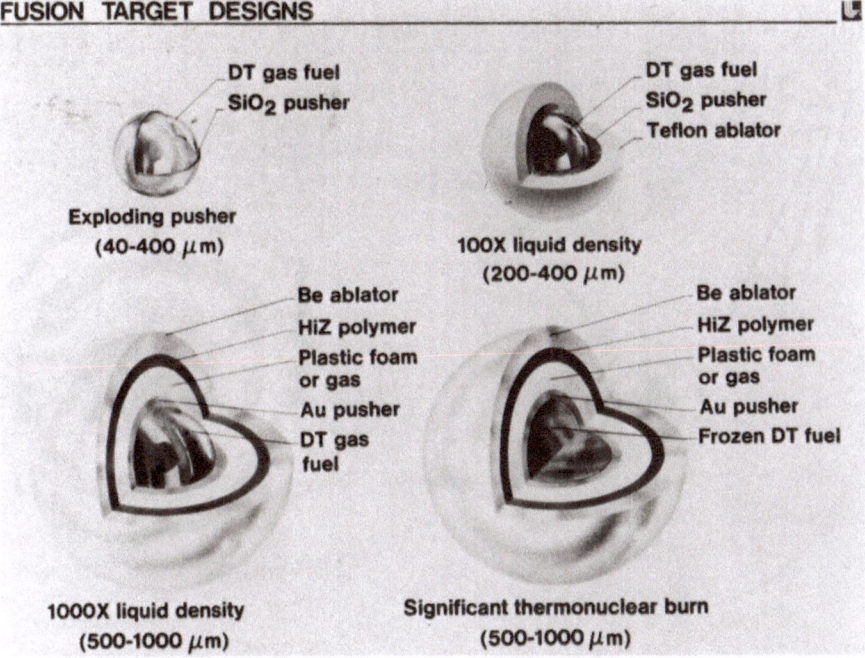

DT gas fuel
SiO2 pusher

Exploding pusher
(40-400 μm)

DT gas fuel
SiO2 pusher
Teflon ablator

100X liquid density
(200-400 μm)

Be ablator
HiZ polymer
Plastic foam
or gas
Au pusher
DT gas
fuel

1000X liquid density
(500-1000 μm)

Be ablator
HiZ polymer
Plastic foam
or gas
Au pusher
Frozen DT fuel

Significant thermonuclear burn
(500-1000 μm)

Fig. 22. Laser fusion target design at Lawrence Livermore Laboratory. Four different pellet configurations are shown.

under way. There is also a large and increasing effort in developing particle beam drivers, with emphasis perhaps tending to ion beams. In several respects, although they are least advanced at present, heavy ions appear to be the most promising. Inertial fusion is a rapidly advancing field, and many new developments may be expected in the 1980s.

6. Future Tokamak Research and JET

The next major milestone in magnetic fusion research is to obtain plasma conditions where $n\tau$ approaches $10^{14} \mathrm{cm^{-3}s}$ and alpha particle heating plays a significant role in the energy balance or even to obtain a net energy gain. To do this it is necessary to increase the confinement time to $\sim 1 \mathrm{s}$. Since $\tau \sim na^2$, large apparatus is required with additional heating from neutral injection or high-frequency waves of 25—100MW for several seconds with currents in the mega-amp range. Since the power necessary to reach given plasma conditions, $e.g.$ ignition, varies as $1/\tau^2$, there is a considerable uncertainty in the requirements.

INITIAL HIGH DENSITY TARGET EXPERIMENTS ON HELIOS

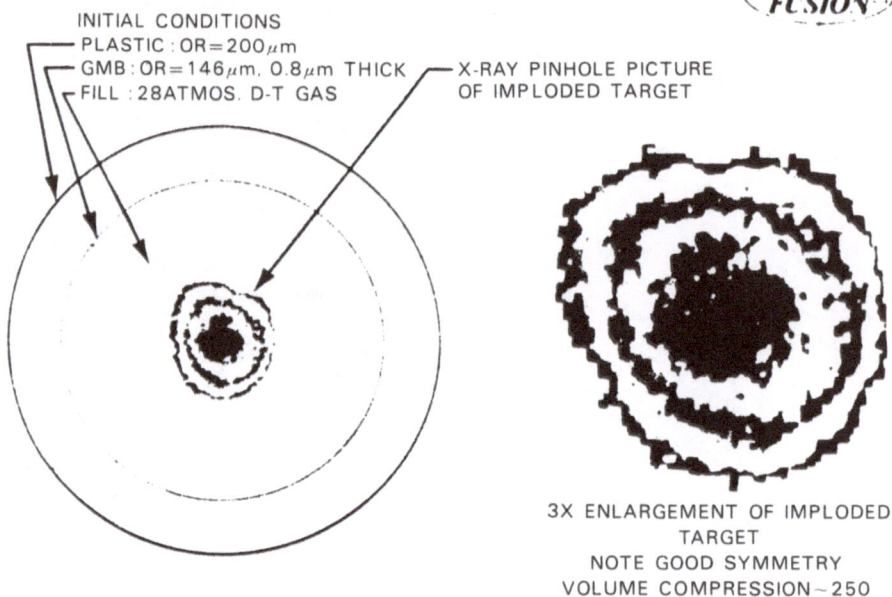

INITIAL CONDITIONS
PLASTIC : OR=200μm
GMB : OR=146μm, 0.8μm THICK
FILL : 28ATMOS. D-T GAS

X-RAY PINHOLE PICTURE
OF IMPLODED TARGET

3X ENLARGEMENT OF IMPLODED
TARGET
NOTE GOOD SYMMETRY
VOLUME COMPRESSION~250

Fig. 23.   X-ray photographs taken through a pin-hole of an imploded pellet using the Helios CO_2 laser at Los Alamos. The left-hand picture illustrates the degree of compression and the right-hand picture is an enlargement of the imploded pellet with a volume compression of about 250.

There are at present six experiments which might approach the ignition regime. JET, JT-60, and TFTR, for all of which construction is well under way; Zephyr (proposed), T-15, and Torus II. Each of these has times in the seconds range, power inputs of tens of megawatts, and currents in the range 1–5MA. The parameters are given in Table 9, together with those of INTOR—the proposed international post-JET experiment presently being designed by an international design study group under the auspices of the IAEA. In Figure 24 the cross-sections of some of these are shown schematically to illustrate the progressive increases in size. The main features of these machines are as follows:

JET (Euratom): The design philosophy is to increase the size while maintaining density and magnetic field similar to large existing machines. The use of non-circular cross section enables higher β to be obtained. It is designed to allow D-T operation with a radioactive assembly and remote control and is expected to have a higher performance than JT-60 or TFTR.

Table 9.   Parameters of large tokamaks.

	R(m)	a(m)	b(m)	B(T)	I(MA)	Flat top time (Bφ) s	Date
JET	3.00	1.25	2.10	2.7	3.8	20.0	1983
				(3.4)	(4.8)		
TFTR	2.65	0.85	0.85	5.2	2.5	1.6	1982
JT-60	3.00	0.95	0.95	4.5	2.7	5.0	1983
Zephyr²*	2.02	0.61	0.61	6.1	2.3	≳5.0	?
					(3.7)		
Torus II	2.15	0.70	0.70	4.5	1.7	∞*	?
T-15	2.40	0.75	0.75	3.5	1–2	∞*	?
INTOR	4.50	1.20	1.90	5.0	4.0	∞*	?

a, b half-width and half-height-respectively, of minor cross-section.
Figures in brackets are extended performance.
* Superconducting Bφ coils.
²* Since this lecture was presented, it has been decided not to build Zephyr.

TFTR (U.S.A.): This machine is based on the two-component plasma concept with neutral injection generating a high-energy plasma component in addition to the thermal target plasma. It will not reach ignition by alpha particle heating but might reach breakeven in which the fusion energy output exceeds the beam energy input (the criteria for this is $n\tau \sim 10^{13}$ cm^{-3} s).

JT-60 (Japan): A machine designed for hydrogen operation to avoid the problems of a radioactive structure. Ignition conditions may be reached but there will be no alpha particle production. It has a diverter.

Zephyr (FRG): Zephyr is designed to operate at high magnetic fields and high densities (like the Alcator series in MIT). Heating by neutral injection and adiabatic compression are planned.

Torus II (France): This is a superconducting tokamak with emphasis on studying radio frequency heating, control of impurities, and operation with long pulses up to 15s.

T-15 (U.S.S.R.): A superconducting machine designed for general studies of plasma conditions close to the reactor regime.

INTOR (IAEA): INTOR is an ignition experiment whose objectives are strongly directed towards technological studies such as blanket design, problems of materials under 14MeV neutron bombardment, tritium handling, reactor design of first wall, and electricity production, in addition to physics studies and the control of burning plasmas.

An artist's impression of JET[63] is shown in Figure 25, and some further parameters are listed in Table 10. It is designed to operate in two phases: standard performance with a current of 3MA and extended performance at 5MA. Because of its large size, the ohmic power density is

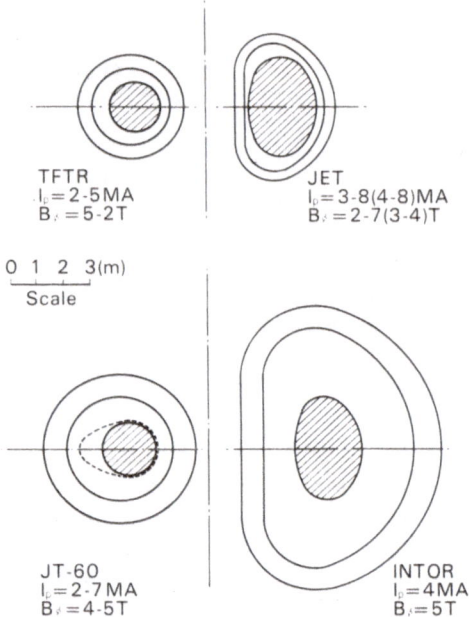

Fig. 24. Cross-sections of the plasma and toroidal field coil arrangement for TFTR, JET, JT–60, and the proposed INTOR design.

relatively small and without injection (assuming Alcator scaling) a temperature of 0.5keV is expected at 3MA and 0.7keV at 5MA. With neutral injection at 10MW, which will be installed after the first year's operation, ion temperatures up to ~ 1.7keV with $\beta \sim 3\%$ are predicted. With extended performance and 25MW injection, ion temperatures ~ 4keV and $\beta \sim 4.5\%$ are predicted. It should be noted that the total power, including all species, is 45MW.

JET is a fully collaborative project within the Euratom Association of the European Community. It is funded approximately: Community 80%, host nation (U.K.) 10%, with 10% equally divided among the partners, who include Switzerland and Sweden from non-Community countries. It is being constructed at Culham Laboratory in the U.K. where it is a separate organization with its own directorate on the same site as the UKAEA Laboratory. Some parts of the machine, *e.g.* neutral injection, are subcontracted to various Euratom laboratories who will provide most of the diagnostics on a similar arrangement. Staff from these laboratories will work on JET.

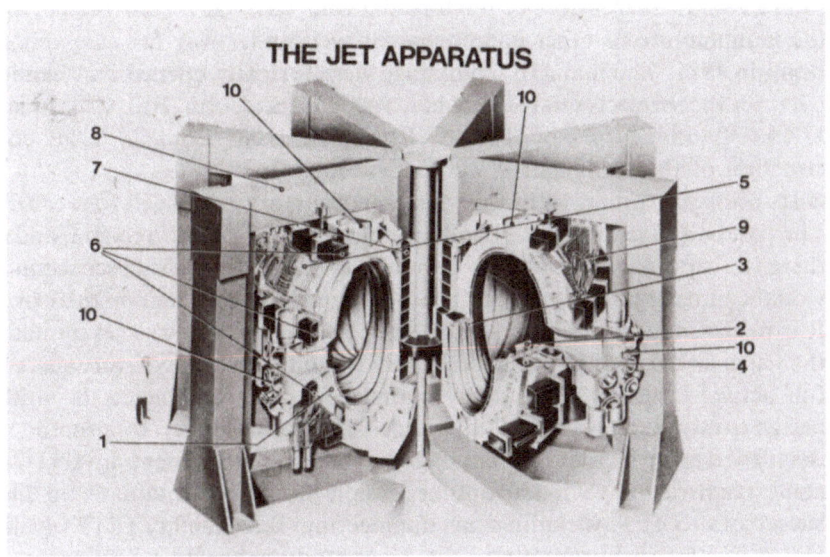

THE JET APPARATUS

Fig. 25. Artist's impression of the JET apparatus (JET Joint Undertaking). (1) vacuum vessel (double walled), (2) limiter defining the outer plasma edge, (3) poloidal protective shields to prevent the plasma touching the vessel, (4) toroidal field magnet of 32 D-shaped coils, (5) mechanical structure, (6) outer poloidal field coils, (7) inner poloidal field coils (primary or magnetizing windings), (8) iron magnetic circuit (core and 8 return sections), (9) water and electrical connections for the toroidal field coils, (10) vertical and radial ports in the vacuum vessel.

Table 10.   Parameters of JET.

R (m)	3
a (m)	1.25
b (m)	2.1
B (T)	2.7 (3.4)
I (MA)	3.8 (4.8)
Rise time (s)	≳ 1
Flat top time (s)	20 (15)
Volt seconds	25 (34)
Plasma volume (m³)	150
Neutral injection energy (keV)	80 (160); $H_2(D_2)$
Neutral injection power (MW) (at full energy)	10 (25)
Toroidal field power supplies (MW)	245 (380)
Stored energy in TF coils (MJ)	940 (1450)

Figures in brackets: extended performance.

The first experiments are scheduled for 1983, and it is presently planned to use D-T in 1986. Most of the toroidal coils have been constructed, and the manufacture of other components is well under way for assembly to begin in 1981. The main JET buildings were formally opened and handed over to the project on 16 October 1980 by Sir John Hill (Chairman, UKAEA) and Professor Teillac (Chairman, JET Council). The construction of the torus hall is well advanced.

It is not yet certain within the fusion community what will follow JET. The cost of a successor to JET will be in the $1,000 − 2,000m region, and so there is a strong incentive to do this on a collaborative basis because individual countries are reluctant to make so large an investment on their own. It is for this reason that the INTOR design study group was set up under the Chairmanship of Dr. S. Mori (JAERI), and it has proved very successful, actively supported by many countries. It has completed the initial design study.[64] Discussions are under way on whether to proceed to detailed design studies. In parallel with their involvement in INTOR, some countries have made outline design studies of alternate possible successors to JET, including the Engineering Test Facility (ETF) in the U.S.A.[65] JET in Euratom and the Tiger study in the U.K.

7. Discussion and Conclusions

During the last 20 years there has been impressive progress in magnetic confinement research. The temperature has risen from about 10^6 °K to about 7×10^7 °K compared with the value of $1 − 2 \times 10^8$ °K needed in a reactor. The confinement time has increased from the microsecond range to almost 1/10 of a second compared with the required value of about 1s. Machines are being constructed, including JT-60 in Japan, on which the conditions necessary to realize a net energy gain from the plasma might be reached within the next decade. In the JET experiment in Europe, which has the highest parameters of those being built and provision to operate in D-T, self-sustaining fusion reactions might be realized. The largest uncertainty is in the value of the confinement time.

Once scientific feasibility is established, the next question is whether an economic fusion reactor can be built. As presently conceived, reactors are complex and likely to be costly, but they are very desirable from an environmental viewpoint. The magnetic fusion community is itself uncertain what to do beyond the next generation of machines. Such a step will be expensive, and international collaboration is attractive, especially for countries which do not feel justified in making the necessary investment on their own. For this reason the INTOR study was set up by the IAEA and has produced the outline design for a yet larger machine. By the turn of the century it should be established whether an economic reactor can be built, and if the confinement physics develops favorably and resources are made available it is just possible that a demonstration reactor could be constructed by this time. The third question, concerning the time scale for the introduction of fusion power (assuming this can be done), is more difficult to answer; one factor upon which it depends is the need for fusion compared with other energy sources, and this will vary from one country to another.

Inertial confinement research grew up rapidly during the last decade, with the development of large lasers and particle beam facilities for pellet compression studies. Lasers with energies extending up to 10 kJ have been used to compress and heat pellets comprising small glass micro-balloons filled with D-T gas. Temperatures of 10^7–10^8 °K have been reached with high compressions but not yet at the very high densities needed for a reactor. During the next decade facilities will be operational which could in principle give a net energy gain for the pellet. These will include large lasers, but the emphasis is changing towards charged particle beam drivers, in particular heavy ions. Many of the general remarks about magnetic fusion also apply to inertial confinement, but in this field there seem to be more unsolved physics problems, especially regarding beam deposition and energy transport in the target.

Fusion research is one of the most challenging and also one of the most difficult tasks mankind has yet undertaken. Since the Department of Nuclear Engineering was founded 20 years ago much progress has been made, and we may look forward to even greater achievements in the next 20 years, which we hope will provide the basis for introducing fusion power on a commercial scale in the next century.

References

1. T. Uchida: paper presented at International Symposium on the 20th anniversary of the Department of Nuclear Engineering, University of Tokyo (1980).
2. Proc. British Nuclear Energy Society Conf. on Nuclear Fusion Reactors, Culham Laboratory (1969), published by Culham Laboratory.
3. R. Hancox and J. T. D. Mitchell: Proc. 6th Int. Conf. on Plasma Physics and Controlled Nuclear Fusion Research, Berchtesgaden, 3, 193 (1977), IAEA Vienna.
4. M. A. Abdou et al.: Proc. 8th Int. Conf. on Plasma Physics and Controlled Nuclear Fusion Research, Brussels, paper El (1980), to be published by IAEA Vienna.
5. B. G. Logan et al.: Proc. 7th Int. Conf. on Plasma Physics and Controlled Nuclear Fusion Research, Innsbruck, 3, 401 (1979), IAEA Vienna.

6. J. L. Emmett: Proc. 9th Euro. Conf. on Controlled Fusion and Plasma Physics, Oxford, p. 396 (1980), published by Culham Laboratory.

7. S. Ido et al.: Proc. 8th Int. Conf. on Plasma Physics and Controlled Nuclear Fusion Research, Brussels, paper E3 (1980), to be published by IAEA Vienna.

8. H. A. Bethe: *Physics Today,* May 1980, p.44.

9. G. Casini: Proc. 9th Euro. Conf. on Controlled Fusion and Plasma Physics, Oxford, 2, 329 (1979), published by Culham Laboratory.

10. R. W. Moir: paper presented at 11th Symp. on Fusion Technology, Oxford (1980), proc. to be published by Culham Laboratory.

11. F. N. Flakus: *Atomic Energy Review,* 13, 587 (1975).

12. J. T. D. Mitchell: Proc. IAEA Workshop on Fusion Design Problems, Culham, p. 517 (1974), published by IAEA Vienna.

13. W. H. Bennett: *Phys. Rev.,* 45, 890 (1934).

14. S. W. Cousins and A. A. Ware: Proc. Phys. Soc. (London) **B 64,** 159 (1951).

15. L. A. Artsimovitch et al.: Proc. 3rd Int. Conf. on Plasma Physics and Controlled Nuclear Fusion Research, Novosibirsk-Nuclear Fusion Supplement, 1, 157 (1969).

16. N. J. Peacock et al.: *Nature,* 224, 448 (1969).

17. N. Fujisawa et al.: Proc. 5th Int. Conf. on Plasma Physics and Controlled Nuclear Fusion Research, Tokyo, 1, 3 (1974), IAEA Vienna.

18. M. Yoshikawa et al.: Proc. 5th Int. Conf. on Plasma Physics and Controlled Nuclear Fusion Research, Tokyo, 1, 17 (1974), IAEA Vienna. H. Maeda et al.: Proc. 6th Int. Conf. on Plasma Physics and Controlled Nuclear Fusion Research, Berchtesgaden, II, 289 (1977), IAEA Vienna.

19. L. A. Artsimovitch: *Nuclear Fusion,* 12, 215 (1972), (tokamak reviews). H. P. Furth: *Nuclear Fusion,* 15, 487 (1975). (*ibid.*) H. A. B. Bodin and B. E. Keen: Reports in Progress in Physics, 40, 1415 (1977), (review of toroidal confinement).

20. J. M. Paul et al.: Proc. 6th Int. Conf. on Plasma Physics and Controlled Nuclear Fusion Research, Berchtesgaden, II, 269 (1977), IAEA Vienna.

21. S. Fairfax et al.: Proc. 8th Int. Conf. on Plasma Physics and Controlled Nuclear Fusion Research, Brussels, paper N–6 (1980), to be published by IAEA Vienna.

22. W. Stodiek et al.: Proc. 8th Int. Conf. on Plasma Physics and Controlled Nuclear Fusion Research, Brussels, paper A–1 (1980), to be published by IAEA Vienna. J. Hosea et al.: Proc. 8th Int. Conf. on Plasma Physics and Controlled Nuclear Fusion Research, Brussels, paper D–5–1 (1980), to be published by IAEA Vienna.

23. A. B. Berlizov et al.: Proc. 8th Int. Conf. on Plasma Physics and Controlled Nuclear Fusion Research, Brussels, paper A–2 (1980), to be published by IAEA Vienna.

24. T. Tamano et al.: Proc. 8th. Int. Conf on Plasma Physics and Controlled Nuclear Fusion Research, Brussels, paper A–3 (1980), to be published by IAEA Vienna.

25. D. M. Mead et al.: Proc. 8th Int. Conf. on Plasma Physics and Controlled

Nuclear Fusion Research, Brussels, paper X–1 (1980), to be published by IAEA Vienna.

26. M. Keilhacker *et al.*: Proc. 8th Int. Conf. on Plasma Physics and Controlled Nuclear Fusion Research, Brussels, paper 0–1 (1980), to be published by IAEA Vienna.

27. M. Murakami *et al.*: Proc. 8th Int. Conf. on Plasma Physics and Controlled Nuclear Fusion Research, Brussels, paper N–1 (1980), to be published by IAEA Vienna. R. C. Isler *et al.*: Proc. 8th Int. Conf. on Plasma Physics and Controlled Nuclear Fusion Research, Brussels, paper A–5 (1980), to be published by IAEA Vienna.

28. M. Suzuki *et al.*: Proc. 8th Int. Conf. on Plasma Physics and Controlled Nuclear Fusion Research, Brussels, paper T–2–3 (1980), to be published by IAEA Vienna.

29. J. R. G. Adam and TFR Group: Proc. 8th Int. Conf. on Plasma Physics and Controlled Nuclear Fusion Research, Brussels, paper D–3 (1980), to be published by IAEA Vienna.

30. H. Kimura *et al.*: Proc. 8th Int. Conf. on Plasma Physics and Controlled Nuclear Fusion Research, Brussels, paper D–5–2 (1980), to be published IAEA Vienna.

31. R. D. Gill *et al.*: Proc. 8th Int. Conf. on Plasma Physics and Controlled Nuclear Fusion Research, Brussels, paper N–4 (1980), to be published by IAEA Vienna.

32. H. A. B. Bodin and A. A. Newton: *Nuclear Fusion*, **20**, 1225 (1980) (review).

33. D. C. Robinson and R. E. King: Proc. 3rd Int. Conf. on Plasma Physics and Controlled Nuclear Fusion Research, Novosibirsk, **1**, 263 (1969), IAEA Vienna.

34. A. Buffa *et al.*: Proc. 8th Int. Conf. on Plasma Physics and Controlled Nuclear Fusion Research, Brussels, paper L–1 (1980), to be published by IAEA Vienna.

35. K. Hirano *et al.*: Proc. 8th Int. Conf. on Plasma Physics and Controlled Nuclear Fusion Research, Brussels, paper L–2–2 (1980), to be published by IAEA Vienna.

36. L. Spitzer: *Phys. Fluids.*, **1**, 253 (1958).

37. D. W. Atkinson *et al.*: Proc. 6th Int. Conf. on Plasma Physics and Controlled Nuclear Fusion Research, Berchtesgaden, **II**, 71 (1977), IAEA Vienna.

38. M. Blaumoser *et al.*: Proc. 6th Int. Conf. on Plasma Physics and Controlled Nuclear Fusion Research, Berchtesgaden, **II**, 71 (1977), IAEA Vienna.

39. D. Akulina *et al.*: Proc. 6th Int. Conf. on Plasma Physics and Controlled Nuclear Fusion Research, Berchtesgaden, **II**, 115 (1977), IAEA Vienna.

40. J. Fujita *et al.*: Proc. 6th Int. Conf. on Plasma Physics and Controlled Nuclear Fusion Research, Berchtesgaden, **II**, 95 (1977), IAEA Vienna.

41. D. V. Bartlett *et al.*: Proc. 8th Int. Conf. on Plasma Physics and Controlled Nuclear Fusion Research, Brussels, paper H–2–2 (1980), to be published by IAEA Vienna.

42. J. Fujita *et al.*: Proc. 8th Int. Conf. on Plasma Physics and Controlled

Nuclear Fusion Research, Brussels, paper H–3–2 (1980), to be published by IAEA Vienna.

43. K. Uo *et al.*: Proc. 8th Int. Conf. on Plasma Physics and Controlled Nuclear Fusion Research, Brussels, paper H–4 (1980), to be published by IAEA Vienna.

44. M. Fujiwara *et al.* and F. W. Baity *et al.*: Proc. 8th Int. Conf. on Plasma Physics and Controlled Nuclear Fusion Research, Brussels, paper BB–4 (1980), to be published by IAEA Vienna.

45. M. N. Bussac *et al.*: Proc. 7th Int. Conf. on Plasma Physics and Controlled Nuclear Fusion Research, Innsbruck, 3, 249 (1979), IAEA Vienna.

46. T. Okawa: General Atomic Co. Report, GA 15561 (1979).

47. A. Mohri *et al.*: Proc. 8th Int. Conf. on Plasma Physics and Controlled Nuclear Fusion Research, Brussels, paper R–5 (1980), to be published by IAEA Vienna.

48. R. F. Post: Proc. 2nd UN Conf. on Peaceful Uses of Atomic Energy, 32, 21 (1958).

49. T. C. Simonen *et al.*: Proc. 8th Int. Conf. on Plasma Physics and Controlled Nuclear Fusion Research, Brussels, paper F–1 (1980), to be published by IAEA Vienna.

50. K. Adati *et al.*: Proc. 8th Int. Conf. on Plasma Physics and Controlled Nuclear Fusion Research, Brussels, paper F–5 (1980), to be published by IAEA. Vienna.

51. J. Nuckolls *et al.*: *Nature*, 239, 139 (1972).

52. G. Sharatis *et al.*: Proc. 5th Int. Conf. on Plasma Physics and Controlled Nuclear Fusion Research, Tokyo, II, 317 (1974), IAEA Vienna.

53. J. L. Emmett *et al.*: Proc. 8th Int. Conf. on Plasma Physics and Controlled Nuclear Fusion Research, Brussels, paper B–1 (1980), to be published by IAEA Vienna.

54. R. B. Perkins *et al.*: Proc. 8th Int. Conf. on Plasma Physics and Controlled Nuclear Fusion Research, Brussels, paper B–2 (1980), to be published by IAEA Vienna.

55. R. Dautray and J. P. Watteau: Proc. 9th Euro. Conf. on Controlled Fusion and Plasma Physics, Oxford, 2, 453 (1979), published by Culham Laboratory.

56. R. Dautray *et al.*: Proc. 8th Int. Conf. on Plasma Physics and Controlled Nuclear Fusion Research, Brussels, paper B–5 (1980), to be published by IAEA Vienna.

57. C. Yamanaka *et al.*: Proc. 8th Int. Conf. on Plasma Physics and Controlled Nuclear Fusion Research, Brussels, paper B–3 (1980), to be published by IAEA Vienna.

58. V. P. Smirnov: Proc. 9th Euro. Conf. on Controlled Fusion and Plasma Physics, Oxford, 2, 473 (1979), published by Culham Laboratory.

59. S. Nakai *et al.*: Proc. 8th Int. Conf. on Plasma Physics and Controlled Nuclear Fusion Research, Brussels, paper P–1 (1980), to be published by IAEA Vienna.

60. G. W. Kuswa: Proc. 8th Int. Conf. on Plasma Physics and Controlled

Nuclear Fusion Research, Brussels, paper P-3 (1980), to be published by IAEA Vienna.

61. G. Cooperstein *et al.*: Proc. 8th Int. Conf. on Plasma Physics and Controlled Nuclear Fusion Research, Brussels, paper P-2 (1980), to be published by IAEA Vienna.

62. R. C. Arnold *et al.*: Proc. 8th Int. Conf. on Plasma Physics and Controlled Nuclear Fusion Research, Brussels, paper P-4-1 (1980), to be published by IAEA Vienna.

63. R. J. Bickerton: Proc. 9th Euro. Conf. on Controlled Fusion and Plasma Physics, Oxford, **II,** 401 (1970), published by Culham Laboratory.

64. The INTOR Report, published IAEA Vienna (1980).

65. W. R. Becraft and P. J. Reardon: Proc. 8th Int. Conf. on Plasma Physics and Controlled Nuclear Fusion Research, Brussels, paper V-1 (1980), to be published by IAEA Vienna.

Japan Fusion Research Activities and Future Plans

Taijiro UCHIDA

1. Introduction

This paper deals with the history, present activity and future prospects of fusion research in Japan.

The basic principles of fusion research and fusion reactors are described elsewhere in this volume by Dr. H. A. B. Bodin.[1] He also reviews world-wide magnetic and inertial confinement studies, presenting the perspective of possible future development in fusion research.

Referring to his paper, I will describe Japanese research history (in Section 2) and the present organization system (in Section 3). Magnetic and inertial fusion research activities with technical developments in Japan are explained in Section 4. International collaboration is reviewed in Section 5, and the future plans and prospects are described in Section 6.

2. History

Japan fusion research started around 1956–58; it was initiated by the famous lecture of Dr. Bhabha at the first Geneva Conference on Peaceful Uses of Atomic Energy in 1955. Like other countries, Japan was excited by his statement that nuclear fusion energy might be released within 20 years.

In many universities, like Osaka, Nagoya, Kyoto, Tokyo, and Nihon Universities, active scientists and engineers gathered and began fusion study in this period, when other countries' experiments—for instance, Zeta[2] in the United Kingdom and Stellarator[3] in the United States—were also in their preliminary stage.

Heated discussion of how to proceed in fusion research and how to concentrate our efforts toward fusion took place on many occasions, and intensive debate took place at the Japan Science Council in 1959. Professor Hideki Yukawa decided that our efforts should be devoted not to building

---

Department of Nuclear Engineering, University of Tokyo, Tokyo, Japan.

a large experimental device but to studying fundamental physics, like plasma physics, in an earlier stage of fusion development.

In accordance with this suggestion oriented to rather basic study, the Institute of Plasma Physics (IPPN) was established by the Ministry of Education, Science and Culture at Nagoya University in 1961, when the First IAEA Conference on Plasma Physics and Controlled Nuclear Fusion Research was held at Satzburg.

Certainly, it seemed that the decision stressing basic study was correct in this phase of research. Japanese contributions to establish fundamental plasma physics were widely recognized in the world, especially in the 1960s.

However, this strategy also seemed to have some negative influence on the promotion of confinement and heating studies of high-temperature plasma, excluding a few cases like the Heliotron[4] project, since few papers on plasma confinement and heating were presented from Japan before around 1968, the time of the Third IAEA Conference held in Novosibirsk.

Outside Japan, the ten years from 1958 to 1968 were spent in looking for the best way to achieve fusion. Worldwide big projects seemed to lose their own ways, except for a few approaches such as mirror machines and multipole devices.

Fortunately, data presented at the Novosibirsk Conference showed that we could cross over the gloomy Bohm barrier which had confronted us who were trying to confine high-temperature plasma stably as long as possible. In particular, the tokamak three (T-3) experiment[5] of the U.S.S.R. demonstrated the longest confinement so that its superiority was recognized even in Western countries.

The ten years from 1969 to 1979 were the tokamak age. In Japan, the first tokamak JFT-II[6] was built in the Japan Atomic Energy Research Institute (JAERI) in 1972 (JFT-I[7] was a hexapole device), and in the universities a tokamak JIPP-T II[8] was also constructed at IPPN in 1974 (JIPP-T I[9] was a stellarator). They are continuing to produce many valuable results.

Similarly, and conjointly with tokamak research, alternative approaches which would be complementary to the tokamak approach were also studied. In Japan, universities took so active an interest in those approaches that many projects, for example the Heliotron in Kyoto University, laser fusion in Osaka University, and the tandem mirror in Tsukuba University, were promoted and started in this period.

JT-60,[10] a large tokamak device like TFTR in the U.S.A., JET in the EC, and T-15 in the U.S.S.R., is under construction at JAERI now, and the next step device may be built by JAERI or a similar national organization. It is apparently recognized that JAERI is responsible for the tokamak path while the universities (including IPPN) are in charge of alternative approaches.

However, there are some differences in the character of research between universities with IPPN and national laboratories with JAERI. The role of the former on science and technology would be, by general consensus, to fill in the fundamental and exploratory research, while the latters' role would be in development and demonstration. Accordingly, even in universities there are several small tokamaks which are for rather basic use and education; similarly, a big toroidal pinch device is under construction in the Electrotechnical Laboratory.

It is widely recognized for fusion research in Japan that universities play an essential role in (i) the fundamental establishment of new disciplines related to fusion research like plasma physics; (ii) contributions to fusion research by experts in various related fields, because fusion is so interdisciplinary; and (iii) education and nurture in long-range fusion research and development.

Figure 1 shows (a) governmental annual budgets given to fusion research and development and (b) their ratios to fission R & D. This shows that the Japanese budget for fusion is about half of that of the U.S. and two-thirds that of the EC, and that the ratio of the magnetic fusion budget to fission's is 10 — 15% of any party for the past several years.

Usually, the Japanese budget does not include salaries.

3. Research Organization in Japan

The Japanese research organization is divided into two parts: laboratories and institutes attached to the Science and Technology Agency, and universities affiliated with the Ministry of Education, Science and Culture. Figure 2 shows the organization system.

The Nuclear Fusion Council was established under the Japan Atomic Energy Commission in 1975. The function of the Council is to coordinate all national projects on fusion research, especially in JAERI and national laboratories. In 1976, the Science Council of the Ministry of Education, Science and Culture also established a Fusion Committee in the Science Council to promote fusion activities on the university side. Both are giving advice and proposals to our government, making strategy on fusion research and its development in Japan.

The sub-committee on fusion research in the Japan Science Council which belongs to the Prime Minister's office was authorized before to recommend the establishment of IPPN to the government, as mentioned

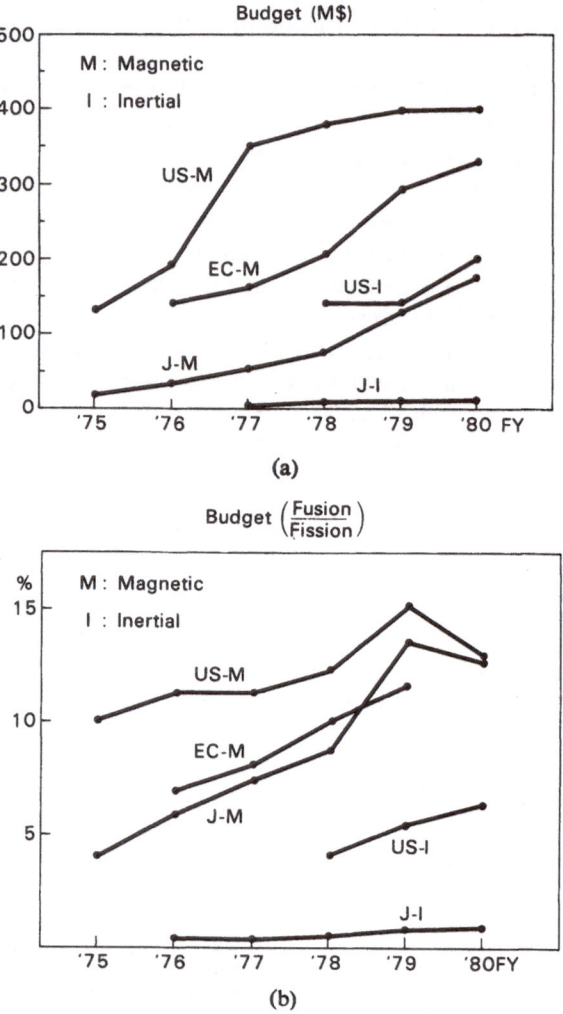

Fig. 1. (a) Governmental annual budgets for fusion R & D.
(b) Ratio of fusion budget to fission's.

above. However, it is rather inactive now. It consists of 210 scholars and professors in all the academic fields and is expected to become powerful again.

Figure 3 shows the internal distribution of Japan's fusion budget. Rapid increases in the budgets for JAERI etc. are due to the construction of JT-60, whose total budget is estimated to be more than one billion dollars.

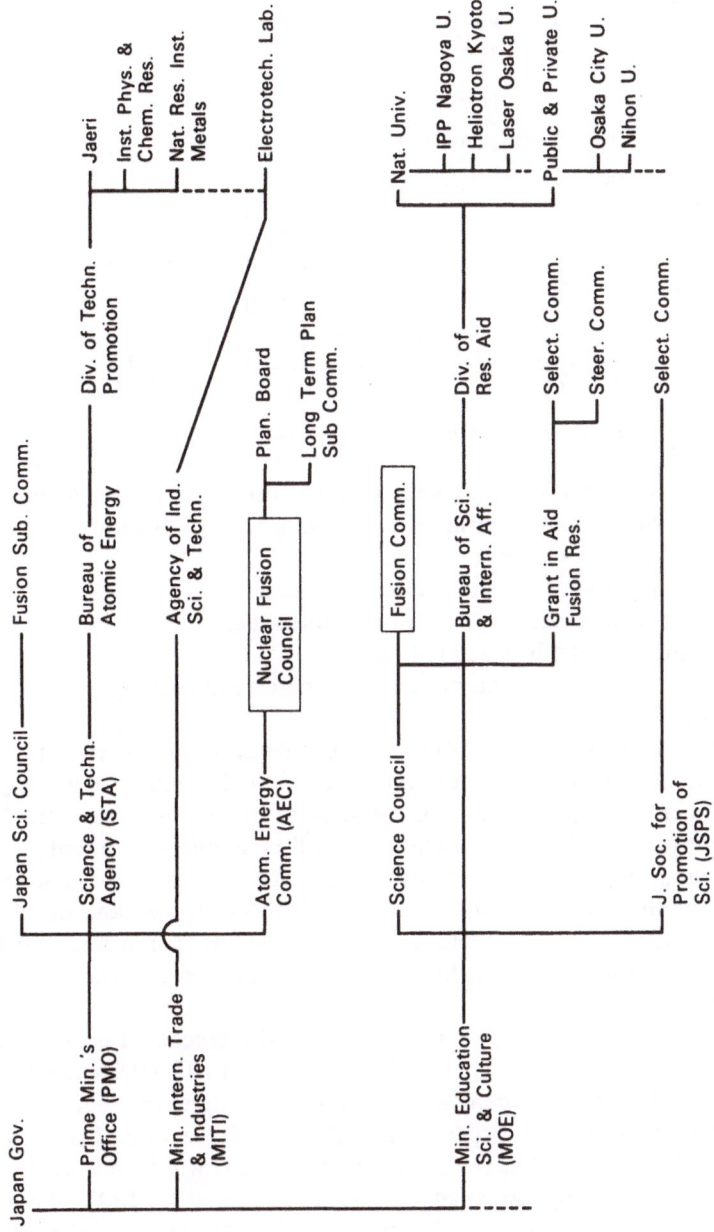

Fig. 2. Organization system of fusion research in Japan (1981).

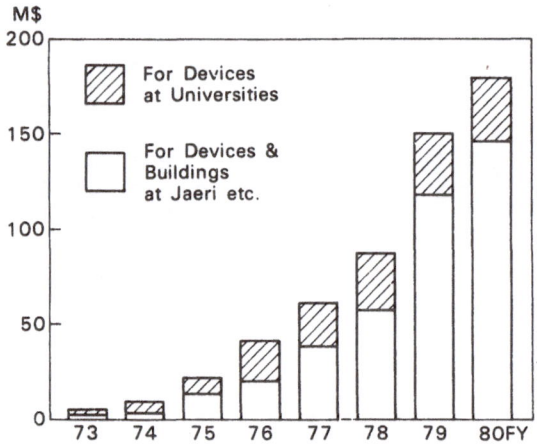

Fig. 3.   Internal distribution of Japan's fusion budget.

In November 1980, the Science Council proposed to the Minister of Education, Science and Culture the following strategy for long-range promotion of fusion research in universities:

1) Set out to study fusion-burning plasma;
2) Select and develop alternative approaches;
3) Promote research in related fields.

The next ten years' research plan in universities will be programmed along these lines.

A grant-in-aid system has begun to subsidize fusion research similarly to other energy-related fields in the 1980 fiscal year. The amount of this grant is not much: 3 million dollars in 1980; however, it can be used freely for many purposes with few restrictions. It will be continued annually.

The Japan Society for Promotion of Science regards fusion research as one of the important interdisciplinary fields to be studied and developed, and is granting a special doctoral fellowship in the fusion research field. JAERI has a similar scholarship for a young research associate who is to work at JAERI.

The long-term planning subcommittee of the Nuclear Fusion Council is discussing how the next step device subsequent to JT-60 should be planned. Here, JT-60 is for the demonstration of a critical plasma, equivalent to a scientific feasibility experiment without D-T reaction. The next machine, called here the Experimental Test Reactor, will be a tokamak like the INTOR[11] being designed at IAEA or like the FED[12] being investigated in the U.S.A. The final report will be presented to the Nuclear Fusion Council in spring 1981.

4. Research Activities

4.1 Tokamaks

There are several tokamaks in Japan. The main devices in operation are JET-II at JAERI and JIPP-T II at IPPN, as mentioned before. Diva,[13] although demounted already at JAERI, produced various notable results. In universities mini-tokamaks—TNT-A[14] and TORIUT-4[15] at the University of Tokyo and TRIAM-I[16] at Kyushu University—are so active that they were evaluated at the 8th IAEA Conference held in Brussels.[17]

The dimensions of those machines are listed in Table 1.

Using the JFT-II machine, several remarkable results were obtained. Figure 4 shows that high beta plasma up to 10% in maximum could be confined stably. The average beta is around 3% when the effective input power is around 1 MW. A lower hybrid heating (LHH) experiment also showed heating efficiency up to 3 eV/kW and demonstrated current sustaining phenomena equivalent to 14 kA, as is seen in Figure 5(a).

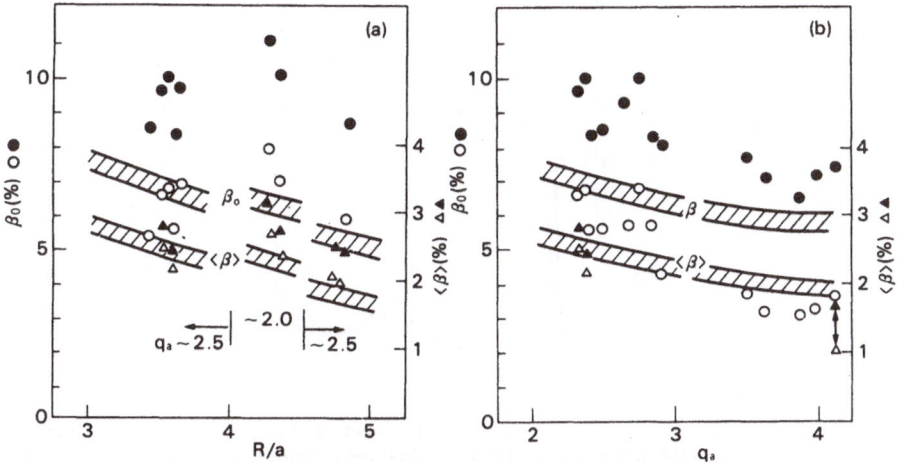

Fig. 4. Beta-value of JFT-II tokamak plasma. β_0 is at plasma center and $\langle\beta\rangle$ is mean value. Two belts show theoretical estimations; circles represent experimental results of β_0, and triangles those of $\langle\beta\rangle$. White shows only the thermal component, while black shows beam and thermal components. Here, q_a means safety factor at plasma edge and R/a is aspect ratio.

An experiment carried out using the JFT-IIa machine, Diva, demonstrated the usefulness of a diverter so obviously that it was recognized that a means for impurity control would be established. Figure 6 is a famous drawing which shows the effective impurity control by Diva. Stable low q operation down to 1.3 and ion cyclotron resonance frequency (ICRF) heat-

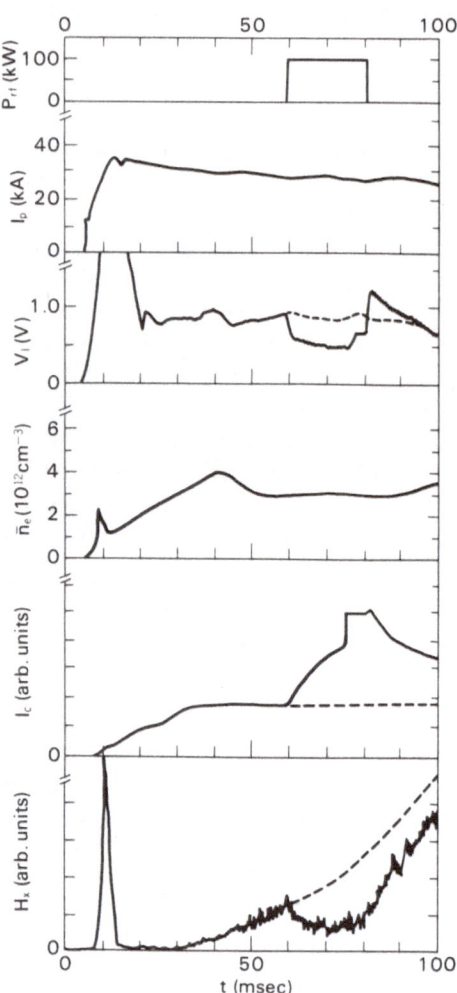

Fig. 5. (a) Time variation of JFT-II plasma parameters of current sustain experiment due to LHH in the case of 1.4 T toroidal field. Solid lines show the case of rf application; broken lines no rf application. From the top, rf power input, plasma current, loop voltage, line-averaged electron density, synchrotron radiation, and hard X-rays emission are presented.

ing efficiency of around 8 eV/kW were also investigated in the Diva experiment.

The large university tokamak JIPP-T II demonstrated skillful position control of the plasma column within 1 cm in radial and in vertical directions and an apparent current increase (Figure 5 (b)) due to LHH similar to that in the JFT-II experiment, as major results obtained so far show.

In spite of its small scale, an experiment with TRIAM I has shown the

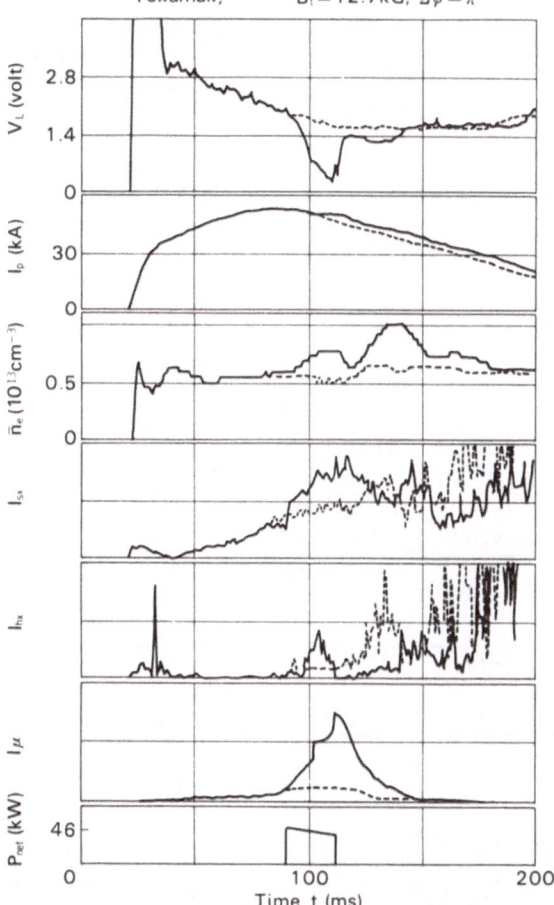

Fig. 5. (b) Time evolution of JIPP-T-II plasma parameters of current sustain experiment due to LHH in the case of 1.3 T toroidal field. The solid (dashed) curves show the wave forms with (without) rf. From the top, loop voltage, plasma current, line-averaged electron density, soft X-rays, hard X-rays, the second harmonic of electron emission, and rf power input are shown.

effective heating due to current-driven turbulence (TH) with pulsive applied toroidal electric field sufficiently in excess of the Dreicer field. The rather stronger toroidal field of 4 T could confine the heated ions, implying neo-classical theory would be applicable. In the TORIUT-4 machine, sawtooth oscillation due to Mirnov instability was observed just before plasma disruption, and stable low q operation was investigated even in the case of non-circular cross-sectional plasma. Figure 7 shows the locus of TORIUT-4 discharge in $1/q_0$-$1/q_a$ diagram. A TNT-A experiment showed stable non-circular cross-sectional plasmas by controlling the plasma current distri-

Table 1. Scale of machines and the plasma parameters of tokamaks in Japan.

	unit	JFT-ll	JIPP T II	DIVA	TNT-A	TORIUT −4	TRIAM −I	WT-II
major radius R	cm	90	91	60	40	30	25.4	40
minor radius a	cm	18–26	17	10	9×13.5	11.5	4.0	9
toroid. field B	T	0.95–1.2	3	0.8 – 2.0	0.37	0.6	4	1.3
plasma curr. I_p	kA	150	160	10 – 80	25	16	47	30
ion temp. T_i	eV	1,000	250–750	70–300 (430 ICRF)	—	33	280 (580)TH	100
ele. temp. T_e	eV	1,200	400–1,300	200–700	250	160	640	300
density n	cm^{-3}	0.25–1.3 ×10^{14}	(1 – 8) ×10^{13}	(2 – 14) ×10^{13}	~5 ×10^{12}	~4 ×10^{13}	2.2 ×10^{14}	(2 – 4) ×10^{13}
particle conf. time τ_p	ms	~15	—	0.5–4	—	<0.5		
energy conf. time τ_E	ms	~15	~14	0.8–5.7	~0.3	0.1 – 0.3	1.8 (0.4 – 0.7) TH	0.4

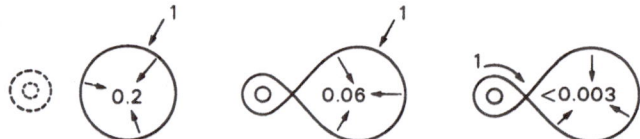

Fig. 6. Schematic figures of impurity influx rate into a hot plasma column from various impurity source positions in Diva.

bution, and the degree of non-circularity reached 1.5. The mini-tokamak WT-II[18] at Kyoto University has demonstrated recently the effectiveness of electron cyclotron heating (ECH) by Gyrotron techniques.

Young students at those small laboratories will be employed in IPPN and JAERI as research associates after graduation.

4.2 Magnetic Alternatives
4.2.1 Heliotron

Heliotron has the longest history of confinement research in Japan. Subsequent to Heliotron A, B, C, P, and D, Heliotron E (Figure 8) was constructed in 1980.[19] It has two helical windings, diameter 4.5 m, and the

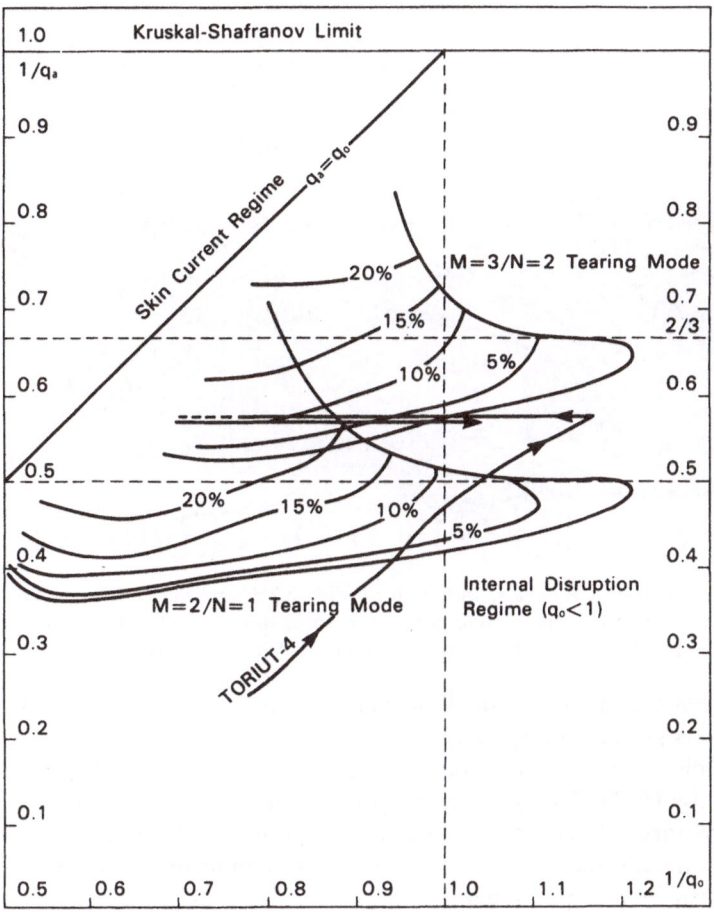

Fig. 7. $1/q_0$–$1/q_a$ diagram. The locus of TORIUT-4 discharge is shown in the figure.

biggest stellarator in the world now. This machine is to be used for the proof of principle experiments. Expected results will be reported in a few years.

4.2.2 Stellarator

The JIPP-T II machine mentioned above can also be used as a stellarator since it has helical windings of $l = 3$.[20] An experiment showed the highest confinement characteristics of plasma heated ohmically in stellarator configuration and proved that the plasma disruption due to large current which is a major concern in tokamak research occurs by no means as far as the helical field is applied.

The non-plane-axis stellarator being studied in Tohoku University[21]

Fig. 8.   Heliotron E device, with major radius 2.2 m and minor 0.2×0.4 m. A pair of helical coils wound along the torus produces 2 T toroidal field at the plasma center. Electric power supply of 330 MVA was prepared.

is believed to be one of the best, but difficult to realize, configuration.

4.2.3   Other Toroidal Systems

A microwave-driven bumpy torus (NBT) experiment is being carried out[22] by IPPN. The confinement principle is the same as that of the Elmo bumpy torus at Oak Ridge National Laboratory (ORNL); however, NBT has an independent historical process of development. Scaling law on confinement will be presented.

At IPPN, a toroidal relativistic electron beam system (SPAC) is also being investigated.[23] A stable plasma ring could be obtained of a few ms duration. This idea would be useful to sustain a poloidal magnetic field without an iron core for the tokamak configuration and is called a compact torus.

Other compact tori have been installed recently in Osaka and Nihon Universities. Preliminary results show stable plasma confinement of about 1/2 ms without tilting instability by using the gun injection technique.[24]

A big toroidal pinch machine with a 10 MJ capacitor bank is now under construction in the Electrotechnical Laboratory (ETL). Its main feature is a kind of screw pinch like the SPICA machine at Utrecht in Holland. A smaller machine at ETL recently showed favorable results in reversed field pinch configuration,[25] similar to the Eta-beta experiment at Padua in Italy.

4.2.4 Open Systems

Tsukuba University is examining a tandem mirror confinement system called the Gamma project, similar to that at Livermore Laboratory (LLL) in the U.S.A. A Gamma 6 experiment demonstrated first the usefulness of ambipolar potential caused by end loss of plasma particles,[26] as is seen in Figure 9. Gamma 10 is now under construction; its scale is around the same as that of TMX at LLL. The main purpose is to recognize the effectiveness of a thermal barrier and axial symmetry in an open system.

A tandem cusp device (RFC-XX) is located in IPPN. An electromagnetic wave applied at both cusp-rings repels escaping light ions like hydrogen back to the cusp, while heavy ions like metal are expelled out of the confinement region. This property was discovered and developed for confinement study by this series of experiments. Recent results[27] have shown similar effect at the spindle cusp region also, as is seen in Figure 10.

4.3 Inertial Confinement

In Japan, inertial confinement fusion is being studied in universities. A research center is located in Osaka University[28] and in a few small laboratories light ion beam (LIB) fusion studies have been started recently.

4.3.1 Laser Fusion

A GEKKO-IV glass laser experiment was carried out at the Institute of Laser Engineering at Osaka University. The peak power is 2 TW with a 4-beam phosphor glass laser, of which the output energy is 1 kJ. Figure 11 shows X-ray images resolved in space and in time. Compression parameters ρR of 3×10^{-3} gcm^{-2} and ρ of 5 gcm^{-3} were obtained with ion temperature of 6 keV.

The main concern at the Osaka Institute is the construction of GEKKO-XII 12-beam glass laser system whose peak power is 40 TW from 20 kJ output energy. This laser will be used for pellet compression experiments to demonstrate significant thermonuclear burn, and the Lawson criterion $n\tau = 10^{14}$ cm^{-3} sec should be achieved. Two modules were already developed and tested. The whole system will be arranged in spring 1983.

A CO_2 laser and relativistic electron beam fusion also are being studied in the Institute. An 8-beam CO_2 laser system of 20 kJ energy is under construction after preliminary experiments with a smaller machine of 1 kJ.

A relativistic electron beam experiment was carried out[29] with a 4 kJ machine whose output parameters are 0.6 MV, 0.2 MA, and 80 ns. Irradiation tests of low Z material resulted in the yield of 10^9 neutrons. New machines of 1.4 MV, 1.4 MA, and 60 ns are being constructed.

Recently, the Technological University of Nagaoka has constructed a 1.0 MV, 0.2 MA, and 50 ns machine for light ion beam fusion.[30] Other universities, Kanazawa University and the Tokyo Institute of Technology,

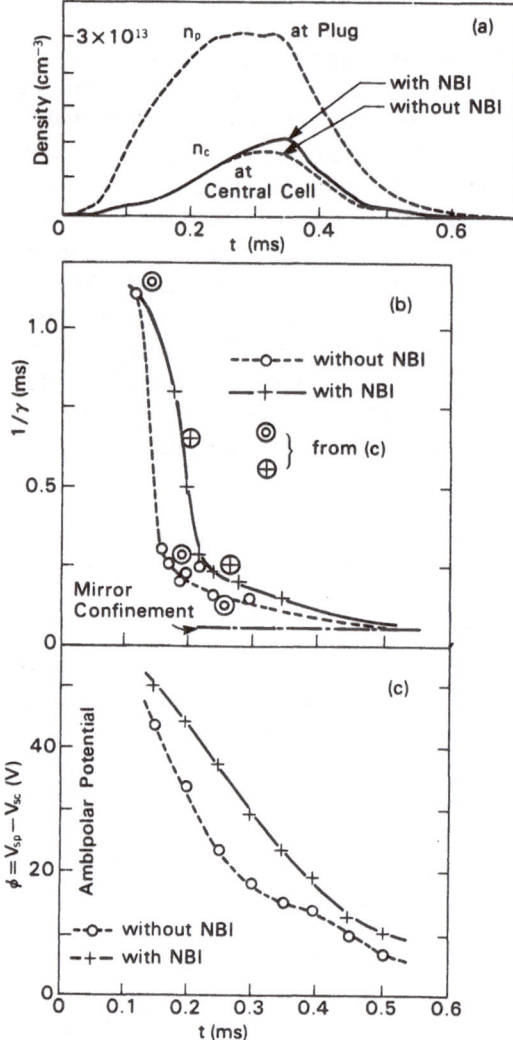

Fig. 9. Time variation of Gamma 6 plasma parameters of ambipolar potential plug experiment due to neutral beam injection (NBI) for ion heating. Solid (broken) lines show the case with (without) NBI.
(a) Densities at central and plug regions.
(b) Inverse of ion loss time.
(c) Ambipolar potential.

have begun to study LIB fusion development in theory and technology.

A new excimer laser has been developed[31] at the University of Electro-communications, and a KrF laser of 1 GW and 10 J is being tested now.

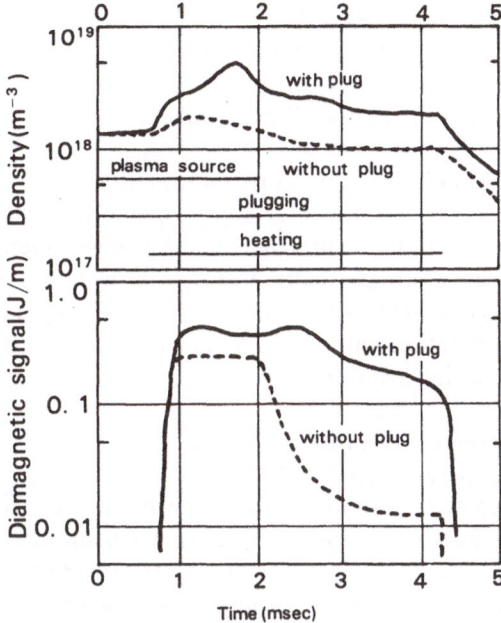

Fig. 10. Time evolution of RFC-XX plasma parameters of rf plugging to ICRF-heated plasma. Solid (dashed) lines show the wave forms with (without) plugging. Upper curves represent density variation and the lower diamagnetic signals.

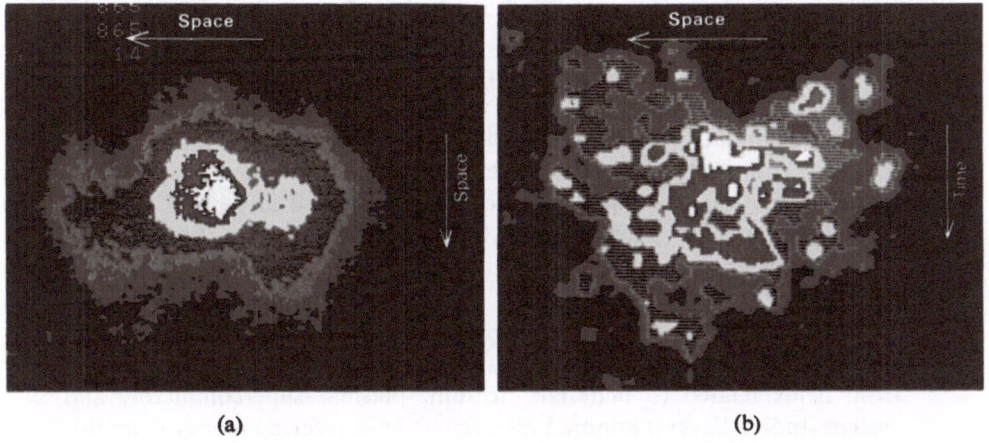

(a)   (b)

Fig. 11. GEKKO experiments. (a) is space-resolved X-ray image where the compression ratio was about 100. (b) is time-space-resolved X-ray image where the implosion speed was order of 10^7 cm/s.

4.4   Theory and Computational Work

In JAERI, a computer cord system TRITON for tokamak research has been established. This is mainly for the fluid model and is being developed and arranged for actual analysis of experimental results and design of machines.

In IPPN, a computer center was organized, and now a FACOM M-200 MP system is set up for all the university people in fusion research. Computer simulation of plasma behavior is being studied intensively.

The Institute for Fusion Theory was established at Hiroshima University, where young research associates are working so actively that international cooperation with the U.S. in progress on plasma theory is mainly centered in this institute.

4.5   Fusion Technology and Reactor Study

4.5.1   Technology Development

At national organizations under the Science and Technology Agency, rather bigger projects on fusion technology are being performed.

In JAERI, a neutral beam injector up to 20 MW, 10 sec is being developed for JT-60. It consists of 14 units whose power is 1.4 MV, 75 kV, and 35 A each. At the same time, JAERI is developing 1 MW Klystron for rf heating of 1.7–2.2 GHz.

For a superconductor magnet, a cluster test device was mounted recently; the cluster test coil (CTC) made of NbTi is 2 m in diameter and will be tested to 7 T, 30 A/mm^2 for 20 MJ stored energy.

Another project, named LCT (Large Coil Task), was started at the initiative of the IEA (International Energy Agency). A large NbTi coil 3.5×4.5 m in the horizontal and vertical dimensions is being wound now. It will be tested to 8T for 800 MJ at ORNL soon along with 5 other coils.

Tritium handling is being arranged in JAERI for process technology and for production technology. A process research center will be built in 1981.

Fusion materials are being studied and developed at JAERI and the National Research Institute for Metals. Stainless steels, high-melting point metals, non-metal-coated metals, and ceramics are to be investigated.

The grant-in-aid mentioned above is being applied to fusion technology study and development in the universities. In 1980, 17 subjects were selected from fields related to materials, tritium, plasma, superconductor, and system studies. Several hundred researchers have received money from this grant.

Tritium research centers have been established at Toyama and Kyushu Universities and Tokyo Institute of Technology. However, the maximum quantity allowed to be treated is about 100 Ci, which is much less than that of a real reactor. Tritum research is a serious problem in Japan.

A CTR Blanket Engineering Research Facility was established in 1974 at the University of Tokyo. Each part of the fusion reactor blanket is being studied using relatively small machines or models. Several doctoral theses on fusion reactor technology are presented every year.

4.5.2 Reactor Study

Tokamak reactor studies are done mainly at JAERI,[32,33] while alternative reactor designs are presented at universities. However, few people work on these, because the important matter for fusion R & D in Japan is how we should proceed to the next stage, especially of the alternatives, rather than reactor design studies. With reference to developments in other countries, this will be decided in a few years.

5. International Cooperation

Figure 12 shows the present status of international fusion collaboration of Japan.

5.1 Multilateral Cooperation

As a member of the United Nations, Japan is working on positive lines for fusion R & D in IAEA. The chairman of the INTOR Workshop is from JAERI, and it is clear that the Japanese government will continue to

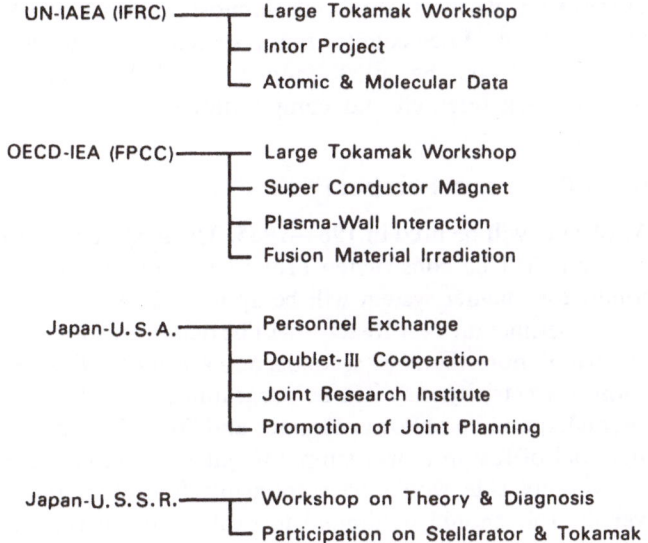

Fig. 12.   International collaboration.

support the workshop up to Phase 2 A, the first half of the design phase. The Japanese representatives are JAERI people only.

Also, as a member of OECD, Japan works along positive lines for IEA fusion activities, participating in (i) the Large Tokamak Workshop, (ii) the LCT project (mentioned before), (iii) the TEXTOR project at Julich in the Federal Republic of Germany for plasma-wall interaction study, and (iv) the FMIT project at Richland in the U.S.A. for fusion material irradiation. We may join (v) cooperation on the RFX project at Culham (U.K.) soon. University people have the initiative on theme (iii) and perhaps on (v).

5.2 Bilateral Cooperation

In 1979, Japan started bilateral cooperation with the U.S. and the U.S.S.R. at the same time.

Personnel exchange with the U.S. is occurring through workshops on special topics, participation in the other's projects, and review tours. Doublet III cooperation is being carried out so smoothly that the remarkable results deserved notice at the 8th IAEA Conference.[34] A joint research institute has branches in each country (the Japanese branch is in IPPN), and intensive research work on plasma theory has begun with long-term exchanges of scientists. As a result of joint planning, cooperation on the RTNS project at LLL is to be realized in 1981, and joint planning on Bumpy Torus and Tandem Mirror researches may be discussed soon.

Workshops on plasma theory and diagnostics with U.S.S.R. were held in 1979 on each side. Each country sent researchers to participate in a stellarator project of the the other country in 1980. Participation in the partner's tokamak research and computational works will follow.

6. Future Plans and Perspectives

JT-60 plasma will be fired in 1984–1985. The next device, if the decision is made soon, will be constructed around the middle of the 1990s. The superconductor magnet system will be applied. The main purpose of the next step Experimental Test Reactor will be demonstration of self-ignition, nuclear burn of more than 100 seconds, and a test of tritium breeding. The generation of electricity may not be programmed. As the subsequent major steps, an Experimental Power Reactor and DEMO are being considered. The final goal of fusion power supply might be materialzed in the 2010s.

The university side would be responsible for alternative approaches. However, its role should be limited to exploration of the feasibility of a reactor due to the fundamental property of universities. In other words, the Ministry of Education, Science and Culture might take care of projects up to a half billion dollars. University activities on fusion R & D in Japan will continue in many fusion-related fields, showing originality and excellence.

References

1. H. A. B. Bodin: paper presented at International Symposium on the 20th Anniversary of the Department of Nuclear Engineering, University of Tokyo (1980).
2. E. P. Butt *et al.*: Proc. 2nd UN Conf. on Peaceful Uses of Atomic Energy, **32**, 42 (1958), IAEA Vienna.
3. T. Coor *et al.*: Proc. 2nd UN Conf. on Peaceful Uses of Atomic Energy, **32**, 201 (1958), IAEA Vienna.
4. K. Uo *et al.*: Proc. 3rd Int. Conf. on Plasma Phys. & Contr. Nucl. Fusion Res., Novosibirsk, **1**, 217 (1969), IAEA Vienna.
5. R. A. Artsimovich *et al.*: Proc. 3rd Int. Conf. on Plasma Phys. & Contr. Nucl. Fusion Res., Novosibirsk, **1**, 157 (1969), IAEA Vienna.
6. N. Suzuki *et al.*: Proc. 8th Int. Conf. on Plasma Phys. & Contr. Nucl. Fusion Res., Brussels, paper T–2–3 (1980), to be published by IAEA Vienna. Yamamoto, T. *et al.*: *Phys. Rev. Lett.*, **45**, 716 (1980).
7. S. Tamura *et al.*: Proc. 4th Int. Conf. on Plasma Phys. & Contr. Nucl. Fusion Res., Madison, **1**, 75 (1971), IAEA Vienna.
8. K. Ohkubo *et al.*: Proc. 4th Top. Conf. on Radio Freq. Plasma Heating, Austin (1981), to be published by Univ. of Texas.
9. K. Miyamoto *et al.*: Proc. 4th Int. Conf. on Plasma Phys. & Contr. Nucl. Fusion Res., Madison, **3**, 93 (1971), IAEA Vienna.
10. Y. Iso *et al.*: Proc. 4th Top. Meeting on Techn. of Contr. Nucl. Fusion, Kings-of-Prussia (1980), to be published by ANS.
11. International Tokamak Reactor: Zero Phase (1980), IAEA Vienna.
12. private communication from Report on USDOE Magnetic Fusion Program (1980), by Fusion Review Panel of the Energy Research Advisory Board.
13. K. Shimomura *et al.*: Proc. 8th Int. Conf. on Plasma Phys. & Contr. Nucl. Fusion Res., Brussels, paper X–2–1 (1980), to be published by IAEA Vienna.
14. K. Miyamoto *et al.*: *Bull. of Amer. Phys. Soc.*, **25**, 901 (1980).
15. M. Kikuchi *et al.*: Proc. Japan-U.S. Workshop on Tokamak Results, ORNL (1980), published by ORNL & USDOE.
16. S. Itoh *et al.*: *Bull. of Amer. Phys. Soc.*, **25**, 901 (1980).
17. K. Toi *et al.*: Proc. 8th Int. Conf. on Plasma Phys. & Contr. Nucl. Fusion Res., Brussels, paper X–4–3 (1980), to be published by IAEA Vienna.
18. S. Tanaka *et al.*: Proc. 2nd Grenoble-Varenna Int. Symp. on Heating in Toroidal Plasmas, Como, (1980), to be published by EPS *et al.*
19. K. Uo *et al.*: Proc. 8th Int. Conf. on Plasma Phys. & Contr. Nucl. Fusion Res., Brussels, paper H–4 (1980), to be published by IAEA Vienna.
20. J. Fujita *et al.*: Proc. 8th Int. Conf. on Plasma Phys. & Contr. Nucl. Fusion Res., Brussels, paper H–3–2 (1980), to be published by IAEA Vienna.

21. Y. Goto *et al.*: Proc. 9th Euro. Conf. on Contr. Fusion & Plasma Phys., Oxford, **2**, 73 (1980), published by Culham Laboratory.
22. M. Fujiwara *et al.*: Proc. 8th Int. Conf. on Plasma Phys. & Contr. Nucl. Fusion Res., Brussels, paper BB–4 (1980), to be published by IAEA Vienna.
23. A. Mohri *et al.*: Proc. 8th Int. Conf. on Plasma Phys. & Contr. Nucl. Fusion Res., Brussels, paper R–5 (1980), to be published by IAEA Vienna.
24. K. Watanabe *et al.*: Proc. Japan-U.S. Workshop on Compact Torus, Osaka (1981), to be published by IPP at Nagoya Univ.
25. Y. Hirano *et al.*: Proc. 8th Int. Conf. on Plasma Phys. & Contr. Nucl. Fusion Res., Brussels, paper L–2–2 (1980), to be published by IAEA Vienna.
26. S. Miyoshi *et al.*: Proc. 8th Int. Conf. on Plasma Phys. & Contr. Nucl. Fusion Res., Brussels, paper F–2–2 (1980), to be published by IAEA Vienna.
27. K. Adati *et al.*: Proc. 8th Int. Conf. on Plasma Phys. & Contr. Nucl. Fusion Res., Brussels, paper F–5 (1980), to be published by IAEA Vienna.
28. H. Azechi *et al.*: Proc. 8th Int. Conf. on Plasma Phys. & Contr. Nucl. Fusion Res., Brussels, paper B–3 (1980), to be published by IAEA Vienna.
29. K. Imasaki *et al.*: Proc. 8th Int. Conf. on Plasma Phys. & Contr. Nucl. Fusion Res., Brussels, paper P–1 (1980), to be published by IAEA Vienna.
30. M. Nakabara: Kakuyugo-Kenkyu [only abstract in English] **45**, 23 (1981), IPP at Nagoya Univ.
31. H. Hara *et al.*: *Japan J. Appl. Phys.*, **19**, L–241 (1980).
32. K. Sako *et al.*: Proc. 5th Int. Conf. on Plasma Phys. & Contr. Nucl. Fusion Res., Tokyo, **3**, 535 (1975), IAEA Vienna. Sako, K. *et al.*: Proc. 6th Int. Conf. on Plasma Phys. & Contr. Nucl. Fusion Res., Berchtesgaden, **3**, 239 (1977), IAEA Vienna.
33. A. Iiyoshi *et al.*: Proc. 5th Int. Conf. on Plasma Phys. & Contr. Nucl. Fusion Res., Tokyo, **3**, 619 (1975), IAEA Vienna.
    Ido, S. *et al.*: Proc. 8th Int. Conf. on Plasma Phys. & Contr. Nucl. Fusion Res., Brussels, paper E–3 (1980), to be published by IAEA Vienna.
34. A. Kitsunezaki *et al.*: *Bull. of Amer. Phys. Soc.*, **25**, 852 (1980).

Discussion: Part V

Dr. M. BENEDICT

Was the gain of 1 % given by Dr. Bodin for the Super Shiva experiment a net gain or ratio of output to input?

Dr. H. A. B. BODIN

The figure of 1 % was the ratio (fusion energy out)/(laser energy into the pellet). It did not include the efficiency of the laser itself, which for neodymium glass is less than 1 %.

Part VI

Nuclear Engineering and Technological Innovation

The Impact of Nuclear Engineering on Technological Improvements in Other Fields

D. G. H. LATZKO

In over 1,000 power reactor operating-years not a single significant accident has occurred due to structural failure. To a large extent this proud and unique record may be ascribed to the achievements of nuclear structural engineering, characterized by the consistent combination of detailed and refined stress analysis, exhaustive compilation of design, offset, and emergency loads, and evaluation of all relevant modes of failure. Many in the audience will recognize the representation of these ingredients given in Figure 1, whose complete version has for many years adorned the back cover of all SMiRT (Structural Mechanics in Reactor Technology) volumes. Together they form the essence of *design by analysis,* the continuous success of which, however, would hardly have been possible without concomitant developments in non-destructive examination (NDE), notably with respect to in-service inspection. The combination of continuously improving methods for flaw detection and description with the advanced design precepts just mentioned has led to the first design code intended to ensure the safe operation of *flawed* structures. This innovation, the importance of which can hardly be overestimated, will figure repeatedly in my presentation.

I see several reasons, each of worldwide validity, for increasing application of these achievements of nuclear structural engineering in non-nuclear technologies. First, nuclear engineering no longer stands out as the energy technology requiring the highest investment per unit capacity. As shown in Figure 2, all of the promising technologies for the 1980s and beyond are at least in the same class of capital intensiveness as nuclear. Nor will nuclear, in my view, remain unique in the intense scrutiny it is rightfully subjected to by regulatory authorities. Events such as the oil gusher from the Ixtoc I oil well in the Gulf of Mexico or the chemical poisoning of the Seveso area in Italy, to name just two examples, cannot in the long run be blanketed by what might perhaps best be termed the Harris-

Delft University of Technology, the Netherlands.

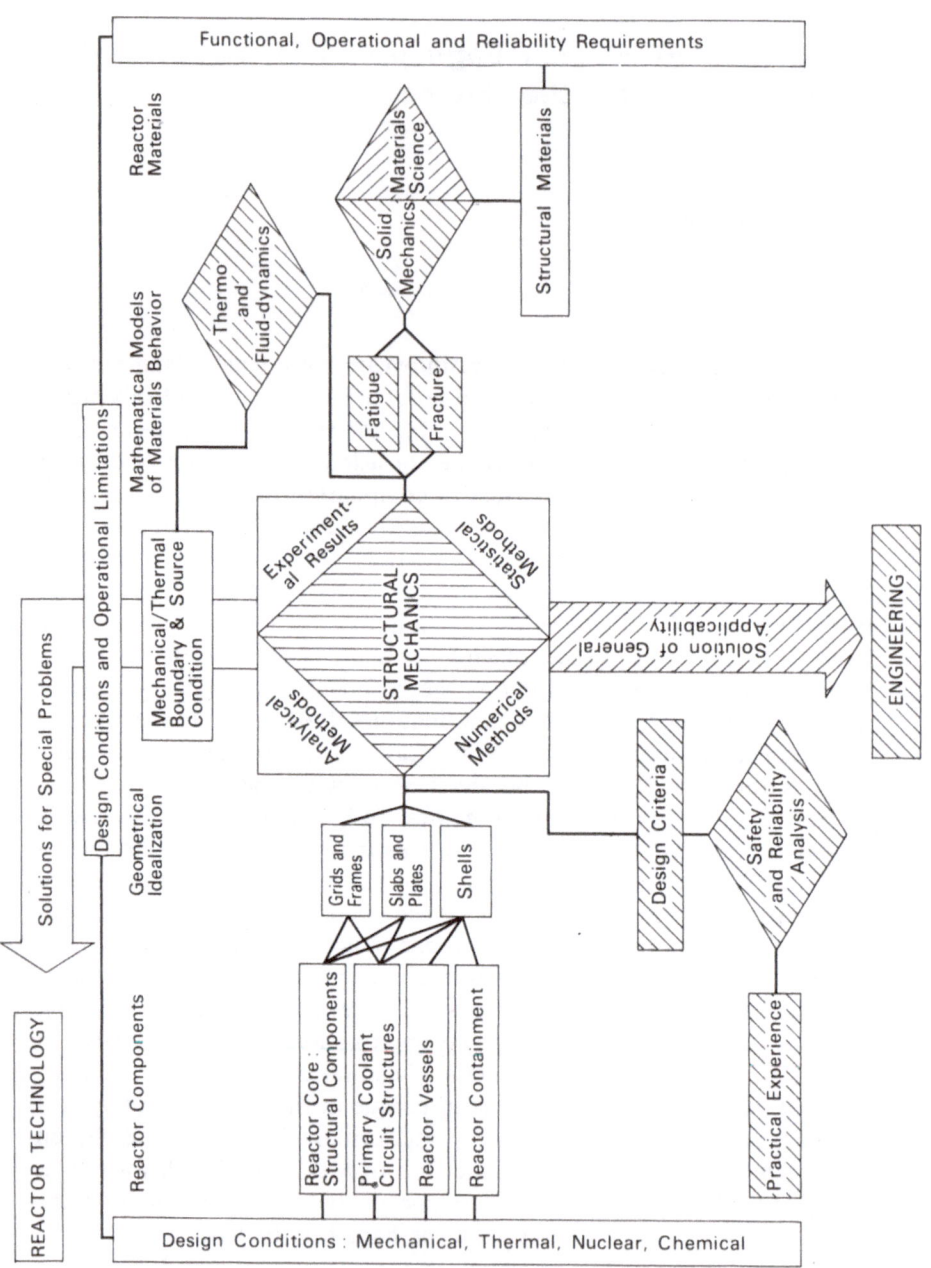

Fig. 1. Characteristics of nuclear structural engineering.

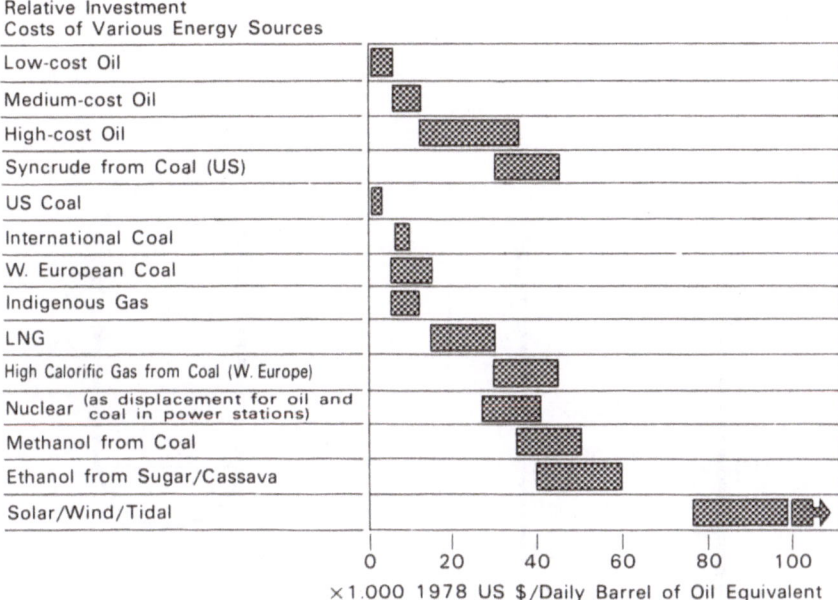

Relative Investment
Costs of Various Energy Sources

Low-cost Oil	
Medium-cost Oil	
High-cost Oil	
Syncrude from Coal (US)	
US Coal	
International Coal	
W. European Coal	
Indigenous Gas	
LNG	
High Calorific Gas from Coal (W. Europe)	
Nuclear (as displacement for oil and coal in power stations)	
Methanol from Coal	
Ethanol from Sugar/Cassava	
Solar/Wind/Tidal	

0   20   40   60   80   100

×1.000 1978 US $/Daily Barrel of Oil Equivalent

Fig. 2.   Relative investment costs of various energy sources.

burg hysteria. On a more rational, though not perhaps more evenhanded, plane, there are the increasingly stringent demands of product liability for equipment manufacturers to be considered. All these aspects, brought about by the need to apply increasingly sophisticated technology in increasingly hostile or vulnerable environments under the scrutiny of an increasingly wary public, emphasize the need to heed the lessons and spread the knowledge of nuclear technology in general and of nuclear structural engineering in particular.

The foremost lesson of nuclear structural engineering is of course the complex of skills and rules constituting the "design by analysis" approach just mentioned. Its importance is illustrated by Figure 3, which I also presented in the course of the Robert D. Wylie Memorial Lecture I had the honor to deliver at the 4th International Conference on Pressure Vessel Technology last May[1]. This repeated use of the same figure to illustrate the same argument within so short an interval is a measure of the importance attached by me to the advantages of this design philosophy. The first of these are the resultant weight savings and cost reductions, shown by the figure as being in the range of 15 to 30%. These weight and cost reductions are due to the detailed stress and failure analysis partly replacing the former so-called "safety factors." They thus represent the substitution of knowledge for ignorance, or if you wish of brains for brawn.

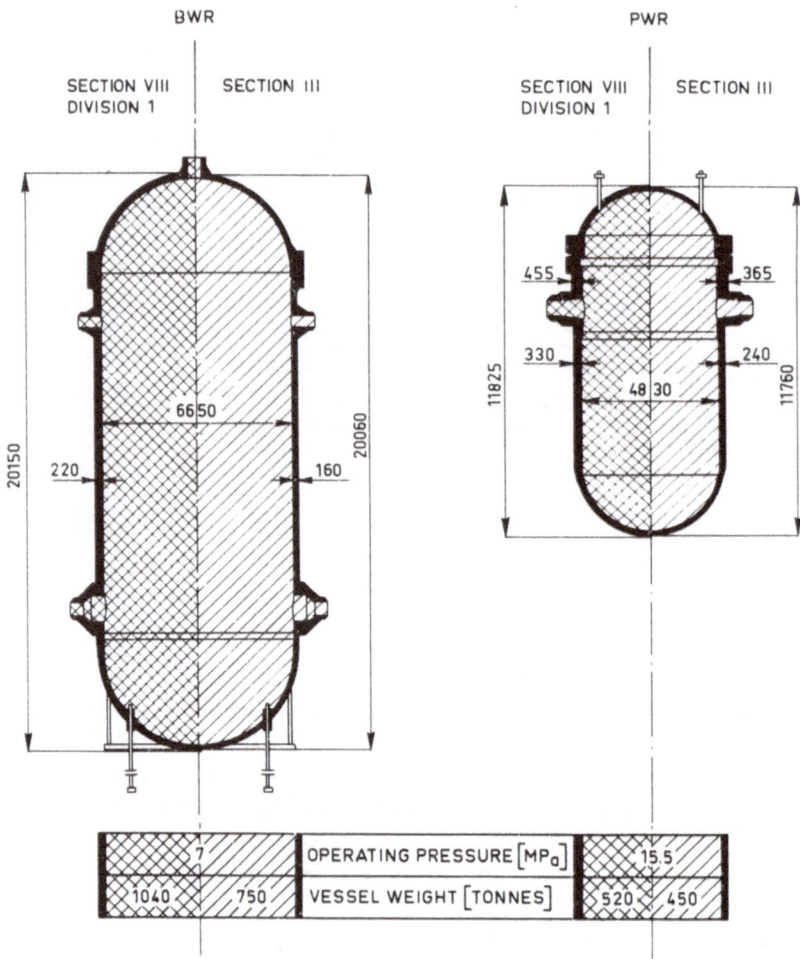

Fig. 3. Savings due to design by analysis.

I have put the words "safety factors" in parentheses to emphasize the fact that the same approach yielding weight reductions also results in increased safety through the systematic evaluation of all relevant modes of failure. It is devoutly to be hoped, for the sake of safety and economy, that the resistances still existing outside the nuclear field against the adoption of codes based on design by analysis will eventually be overcome. The tendency towards design by analysis in fields other than nuclear may be intensified in the near future by the coming of age of computer-aided design, as digital equipment continues its triumphal procession through design offices all over the world. The close logical link between computer-aided

design (CAD) and computer-aided manufacture (CAM) will then undoubtedly increase the rate of introduction of the latter. This in turn will
facilitate, economically and administratively, the transfer of nuclear-type
quality control and quality assurance procedures to many non-nuclear
manufacturing operations.

In talking about manufacturing it is worthwhile noting that considerable benefit from nuclear technology is already being reaped or is definitely
foreseen in a number of areas and ways. One such area is the manufacture
of reactor pressure vessels for coal conversion plants. Gasifier vessels up to
10 m in diameter and over 30 m in height are under serious consideration
for full-scale plants. Designed for pressures between 3 and 7 MPa, these
vessels will have shell thicknesses over 200 mm. While dissolver reactor
vessels for coal liquefaction will be much smaller, typically some 4 m in
diameter, their much higher design pressures may require wall thicknesses
of about 400 mm. In both cases the corrosive and hydrogen-rich environment will require the use of corrosion-resistant cladding, much the same as
in LWR nuclear vessels though for somewhat different reasons.

Manufacturers now investigating the feasibility of such huge vessels will
for example be able to utilize the impressive progress in forging size and
quality achieved in the production of nuclear pressure vessels.[2] They are
also bound to benefit from the various recent developments in heavy section welding techniques. An example of the potential gains from such developments are shown in Figures 4 and 5. The former illustrates the
reduction in weld length attained by the use of large forgings. The latter indicates potential cost savings in the manufacture of large heavy section
vessels from utilizing the narrow gap welding technique.[3,4] The purpose
in showing this latter figure is not so much to emphasize the advantages of

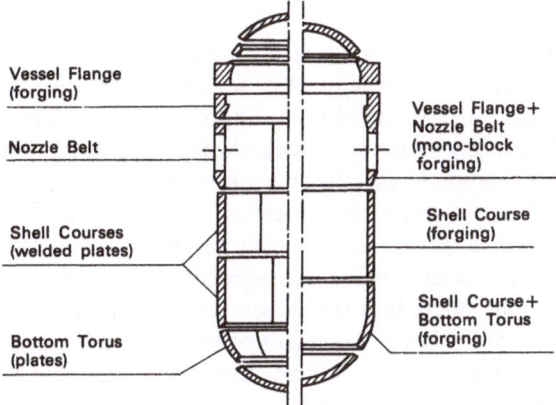

Vessel Flange
(forging)

Nozzle Belt

Vessel Flange +
Nozzle Belt
(mono-block
forging)

Shell Courses
(welded plates)

Shell Course
(forging)

Shell Course +
Bottom Torus
(forging)

Bottom Torus
(plates)

Fig. 4. Impact of forging developments on pressure vessel ISI requirements.

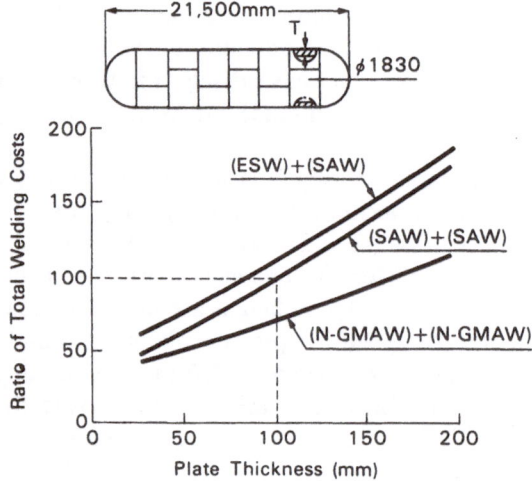

Fig. 5.   Economic impact of new welding processes.

one particular welding technique as to indicate the innovative impulses emanating from nuclear manufacturing needs. Similar impulses, both qualitative and economic in character, have led to significant improvements in weld cladding of vessel inner surfaces.[5] While the quantitative process data of nuclear reactor vessel cladding are not applicable to the higher alloy Cr-Mo steels of interest for coal conversion reactors, the technological know-how thus obtained largely is. It is worth mentioning in passing that the foremost developments in all these areas have occurred in Japan. Needless to say, these manufacturing developments have been attended with significant improvements in quality control techniques, the importance of which to non-nuclear fields will be discussed later in this lecture.

Earlier I mentioned the emergence, in nuclear engineering, of the first code dealing with the assessment of flaws and hence with the safe operation of defect-bearing structures.[6] An impressive and still growing body of experimental data on static and dynamic fracture toughness and fatigue crack propagation data in various environments, backed by theoretical and numerical analyses for all kinds of defect shapes and locations, underlies this code. This solid, but sometimes overconservative, base of linear elastic fracture mechanics is now being extended into the realm of elasto-plastic materials behavior, including the effects of slow stable crack growth.

It is obvious that this vast reservoir of knowledge might have remained closed to practice had not similar advances been achieved in the areas of flaw detection and sizes. I hope to explain below the reasons for using "similar" rather than "equal" in the preceding sentence. As it is, this code

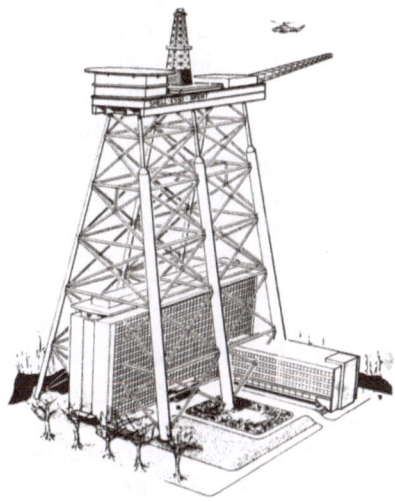

Fig. 6. Artist's impression of steel offshore platform.

should be highly beneficial in all areas combining significant safety risks with high capital investment. One case in point are the steel structures carrying offshore petroleum and gas production platforms. Figure 6 gives an impression of the size of such a structure by comparing it to the 14–story main office building of the Royal Dutch/Shell group. The mass ($\approx$ 27 M kg) and present cost ($\approx$ 300 M \$) are as impressive as the sight, and so are the potential consequences of a serious structural failure. This makes it of the utmost importance to obtain for these structures the kind of detailed knowledge of load spectra and resulting stresses that are typical of Class I nuclear components.

The loading of offshore structures differs essentially from that of land-based plants by both the predominance and the completely stochastic nature of the alternating load component. Furthermore, this predominance is mostly due to the *impact* loads produced by the waves, requiring a truly dynamic analysis.

The three major elements of any fatigue analysis are local stress history, local stress concentration factors, and crack initiation and propagation under alternating stresses in the material of concern.

The stresses are obtained from the postulated wave spectra by multiplying the latter by (the square of) the so-called stress-transfer functions. The generation of these stress-transfer functions, defining the ratio of response amplitude to wave amplitude, thus becomes an essential item in the probabilistic fatigue analysis of offshore structures, The solutions used at

present for this fluid-structure interaction problem (cf. *e.g.* (7)) might benefit from the procedures developed and the vast amount of experimental
evidence gathered in the nuclear field, notably in such areas as core barrel
and steam generator divider plate behavior under post-LOCA pressure
waves, where linear structural response and single-phase flow are usually
part of the underlying assumptions (cf. *e.g.* (8)). An organized exchange
of knowledge between the respective specialists has not, to my knowledge,
taken place up till now and appears overdue.

Another area of common interest, limited in scope but not in importance, concerns the stress concentration factors and fatigue resistance of T
junctions (and other tubular joints such as Y joints) under external loads.
The excellent review recently prepared by Battelle-Columbus Laboratories[9] reveals i.a. that the data complied on the *static strength* of T joints
under in-plane bending loads are neither inclusive of or compared to test
results obtained in connection to nuclear applications, *e.g.*(10). The same
holds for the stress concentration data presented in the same report, despite the fact that a wealth of information on this subject is now available
from nuclear piping studies. Another insight to be gleaned from the BCL
report, and one that may come as a surprise to engineers with a nuclear
background, concerns the definition of what are called the "hot spot"
stresses, *i.e.* the local stress peaks underlying the aforementioned fatigue
evaluation. Though lacking a clear definition, they apparently do not include the stress peaks due to the geometrical discontinuities at the toes of
the fillet welds (cf. *e.g.*(11)). Again it would seem that an exchange of views
with nuclear designers might prove beneficial for clarifying a point of the
utmost importance to the safety of steel offshore structures.

The stochastic nature of the loads on offshore structures leads to probabilistic fatigue analyses, logically to be extended to probabilistic fracture
mechanics structural assessments in case of existing or postulated defects.
Here again much ground has already been covered in the nuclear field, as
borne out *e.g.* by the two International Seminars on Structural Reliability
of Mechanical Components & Sub-assemblies of Nuclear Power Plants,
held in conjunction with SMiRT-4 and -5, respectively. As apparent from
the seminar proceedings, the application of probabilistic methods in nuclear engineering now extends beyond fracture mechanics and flaw assessment to cover seismic safety, systems reliability, and risk assessment. Such
extensions appear likewise feasible in such large-scale, potentially high-
risk activities as offshore oil production, LNG transport and storage, or
coal gasification, to name just a few.

Preparations for the large-scale introduction of coal conversion cannot
be limited to design studies of future equipment, but will have to include
the timely qualification of structural materials as well. In view of the long

duration of at least part of the relevant tests this is already now an urgent matter. While an impressive amount of mechanical behavior data has by now been gathered for $2\frac{1}{4}$ Cr-1 Mo steel (ASTM A387-74A-Gr-22-Cl.2) mainly due to its use in hydrocracking and other petroleum refinery vessels, there remains a need for steels superior to the $2\frac{1}{4}$ Cr-1 Mo variety in hardenability and at least comparable to this steel in elevated temperature strength. While most of the alloys so far investigated for use in future LMFBR steam generators will probably remain too expensive to apply as pressure boundary materials in reactor vessels, they may well be of interest for other applications in coal conversion plants. In addition, at least some part of the considerable amount of metallurgical data gathered for these steels may be of a sufficiently general nature to be useful in the continuing search for improved steels for coal conversion.

While the degree of applicability of nuclear know-how may be debatable in this area, no doubt exists on the applicability of the fracture toughness characterization methods now in general use for nuclear pressure boundary materials. It is interesting to note how K_{IC} and J_{IC} have all but displaced CVN and drop weight test values in recent publications on the evaluation of candidate steels for coal conversion reactors. The relevance of the fracture mechanics approach to toughness characterization in this field of technology is enhanced by the overwhelming importance of hydrogen-induced cracking. For quantification of creep resistance, with its inherently time-consuming and expensive tests for both base material and weldments, the same remarks apply that were made earlier on the applicability of LMFBR data. In addition it is to be expected that the information generated in LMFBR programs on the relation between creep strength properties under uniaxial and biaxial loading will be directly relevant for the steels under consideration for coal conversion equipment.

It is now time to focus our attention on a topic repeatedly touched upon in the foregoing, viz., the developments in non-destructive examination stimulated by nuclear power requirements. In my view their origin is to be found in the shift in design philosophy evidenced by the aforementioned flaw assessment procedures, from a zero defects approach to one which assumes the presence of flaws and designs structures to tolerate them. Prior to this change, the main emphasis of research and development in non-destructive testing by ultrasonics, the method most widely used in nuclear engineering, was on improved sensitivity and reliability of flaw detection.* Present efforts are mainly concerned with obtaining data on

---

* It should be noted that one of the most important programs directed towards improved ultrasonic flaw assessment capability, the Interdisciplinary Program for Quantitative NDE started in 1974 in the United States, with the Rockwell International Science Center as main contractor, had its origin in aerospace rather than in nuclear requirements.

flaw size and shape. Unfortunately, present-day ultrasonic examination is a comparative method, requiring extensive operator experience to relate instrument output to defect size. Hence, while much has been achieved over the past decade, much still remains to be done. A sobering reminder of this extent resulted from the so-called PISC (for Plate Inspection Steering Committee) program. This cooperative program, set up within the framework of the OECD Committee on the Safety of Nuclear Installations, was designed to evaluate the capability of various ultrasonic test procedures to detect, size, and locate artificial, fabrication-induced defects in a heavy section steel plate successively placed at the disposal of 34 European testing institutions. Both standard (ASME Code Section XI) and advanced, high-sensitivity techniques were employed. The results were disappointing in that a significant part of the defects up to about 10 mm remained undetected by a majority of the participants when using standard techniques. Significantly improved results were obtained, however, when using advanced techniques.

In view of the essential need for reliable flaw detection and sizing as a basis for the rational flaw assessment approach mentioned earlier as one of the attainments of nuclear technology, a sizeable effort, sponsored by a variety of institutions in a number of countries, is therefore under way for the further development of such advanced techniques. These efforts take two distinct directions. One aims at improved *measuring* methods. Examples of this direction are ultrasonic holography, a technique which owes much to optical developments based on lasers, and ultrasonic diffraction, in which transit time rather than reflected pulse amplitude is the measured variable. The other, and in my view more generally important, direction is based on increased sophistication in *data processing*. Its general importance derives from the fact that data processing may be, and usually is, spatially separated from data collection, and hence appears eminently suitable to the examination of structures operating in hostile and inaccessible environments.

In the Wylie Memorial Lecture I gave several examples of the use of digital circuitry both for signal generation and for signal processing. Rather than repeating these examples here I should like to reiterate my conclusion, viz., that the mating of advanced computer usage to refined measuring techniques is essential to the further development of NDE. In this trend I again perceive nuclear energy as providing the "market pull," with many advanced technologies as potential beneficiaries. The foremost of these is perhaps offshore technology, where accessibility *and* noise problems are unique in their severity. The rewards, however, are also high, considering that the detection capability of the visual and magnetic particle inspection methods now in almost exclusive use is limited to *surface* defects. Increased awareness within the organizations engaged in and responsible for offshore structural inspection of the latest developments in signal processing should stimulate their interest in ultrasonics. The present self-contained, diver-held, recordless units might then in due course be replaced by remotely operated, platform-based units incorporating the latest developments in signal processing.

The need for such equipment in the 1980s appears at least as pressing in the offshore as in the nuclear field, as oil and gas production extend to ever greater depths, where remotely controlled vehicles become the obvious replacement for human divers. Figure 8 shows an example of such a remotely controlled vehicle, the likes of which are being built in the Netherlands for payloads of 300 kg and a maximum operating depth of 1,000 m. The remote operation of such vehicles is facilitated by the use of optical fibers for their umbilicals. This development, resulting in transmission lines impervious to electrical interference, also appears of essential importance to U.S. scanning at significant distances from the data processing equipment.

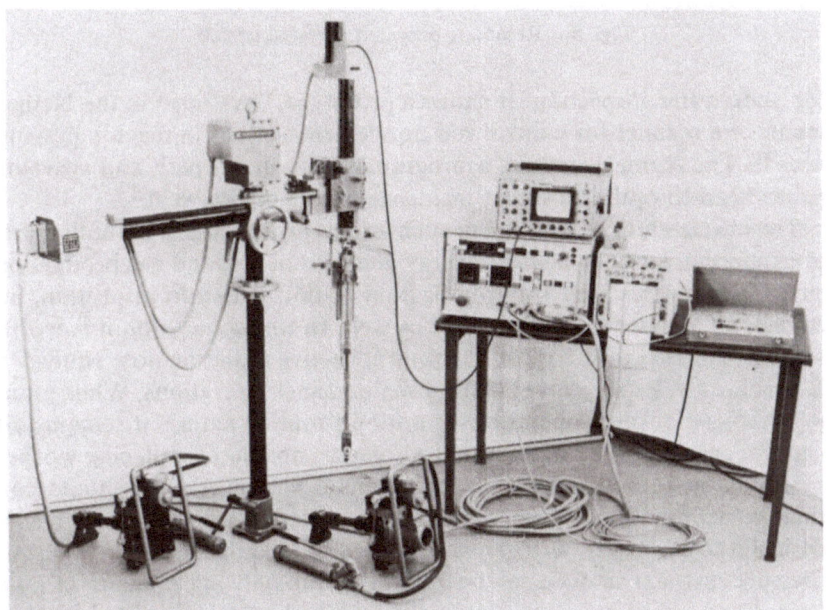

Fig. 7.   Remotely operated ultrasonic scanner.

Figure 7 shows an example of the kind of remotely operated equipment originating from the nuclear field that could initiate similar developments

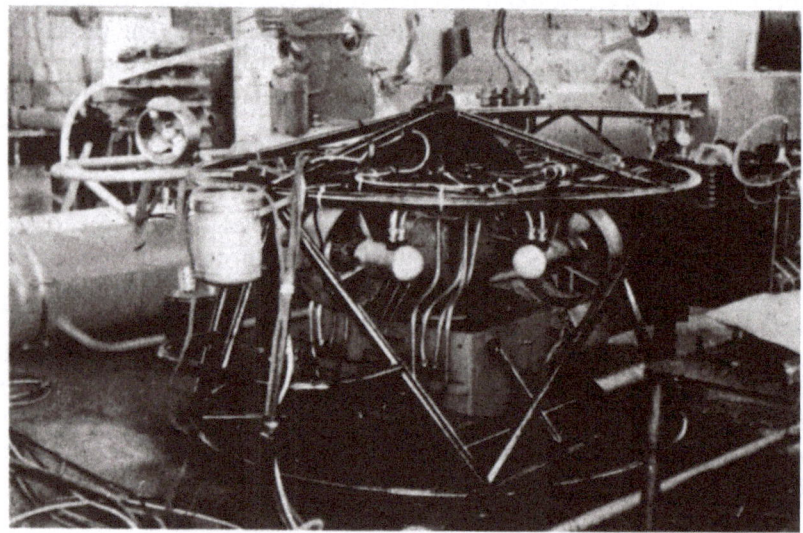

Fig. 8.   Remotely operated deep-sea vehicle.

for underwater inspection. It shows a prototype, developed in the Netherlands, of a scanner for control rod nozzle penetrations in reactor pressure
vessels. The scanner features a programmed scanning path and swiveling
probe head to optimize sound incidence on the nozzle weld.

The robotization of equipment such as that just discussed is another area
of common interest to nuclear energy and offshore oil and gas production.
In both cases the need for robotization is not limited to inspection, but
extends to maintenance operations as well. In this connection it is worthwhile noting that over 80 % of the total collective radiation dose equivalent
in existing LWRs is received during maintenance operations. While granting that part of these operations is not of a routine nature, it remains difficult to accept in the long run that a society capable of replacing workers
by robots in automobile plants for purely economic reasons should continue to expose human beings to ionizing radiation where robots may be
designed to take over. At present it is hard to tell whether the hydrostatic
pressure of water in deep-sea operations or the political pressure of labor
unions and public opinion in nuclear plant maintenance will be predomi-
nant in bringing about the expansion of robot technology from mind-dulling
to downright hostile environments. In either case it would come as no surprise to me if Japan, after leading the world in the former field of robotization, were eventually to take the lead in the latter as well. The world

market, estimated at some 260 million dollars for underwater inspection alone, would certainly seem to warrant a sizeable development effort.

While discussing remotely operated inspection equipment, it seems proper to pay tribute to the general reliability of the instrumentation and mechanical equipment operating in the high-radiation areas of nuclear installations. While it is generally true that most high-technology achievements are of litle or no use to the two-thirds of humanity living in developing countries, the near future may show this to be different for these components. Many of the design and manufacturing principles underlying their prolonged successful operation with a minimum of maintenance should also be applicable to instrumentation and equipment for service under conditions where maintenance is limited by available skills rather than accessibility. As a personal aside I should like to add that the realization of such a fact is just one of the many small ways in which we can stop relegating our materially less endowed fellow men to the role of poor cousins.

A strong need for advanced, reliable NDE techniques also exists in coal conversion technology, as illustrated by the following three examples. The early detection of wall thickness changes due to erosion is essential to the long-term reliability of coal slurry piping and heating systems. Ultrasonic monitoring appears attractive, provided the detectors are capable of operating at temperatures up to 700 °C. The solution to this problem now under investigation, viz., metallic waveguides, requires special electronic circuitry to account for the loss in pulse echoes in these waveguides. Again, a viable solution is seen to depend on adequate signal processing.

Another highly critical group of components in coal processing are valves, viz., their tendency to leak as a consequence of erosion and wear. One approach to leak detection, presently under development at ANL for liquid slurries, is based on flow-induced acoustic energy and tries to correlate leakage rates with acoustic energy as measured at the valve body.

Finally, the intention of detecting thermal stress-induced cracks in the reactor vessel refractory liner during gasifier startup and shutdown by acoustic emission should encounter similar signal-to-noise problems as those found in the earlier tentative applications to nuclear reactor vessels. Information obtained in the course of the latter development programs is undoubtedly relevant and may prove crucial to the operational reliability of future coal conversion reactors.

All aspects of nuclear engineering discussed so far belong to the realm of the inanimate. It is time to look for some lessons nuclear energy has taught us in the field of human behavior.

As early as 1976, a 400-page report on human factors in the nuclear control room was published,[12] summarizing a study sponsored by the Electric Power Research Institute (EPRI). This study identified a host of human factor problems and made suggestions for needed improvements. Yet little heed seems to have been paid to these findings and suggestions until the Kemeny Commission's indictment in the aftermath of Three Mile Island[13]:

The equipment was sufficiently good that, except for human failures, the major accident at Three Mile Island would have been a minor incident. But, wherever we looked, we found problems with the human beings who operate the plant.

A flurry of activities in human factors engineering followed in the wake of TMI, resulting i.a. in a Human Factors Handbook soon to be published by the USNRC, after review by an IEEE Workshop.[14] It would be short-sighted, to say the least, if application of the results of these activities were to be restricted to the nuclear field. It seems highly likely in fact that several serious nautical disasters of the recent past would not have occurred had responsible crew members been subjected to the kind of simulator training now almost generally required for nuclear plant operators. Similar general benefits can be foreseen for the results of studies now under way on control room layout, visual display types, operator stress and fatigue abatement, etc. One would almost expect to see companies operating large process plants standing in line to co-sponsor such studies.

While on the subject of human behavior, it is tempting to dwell upon the motives and perceptions that have turned nuclear energy into a menacing symbol of "future shock" in the eyes of sometimes sizeable parts of the population in many free world countries. Being an engineer unqualified in sociology I shall resist this temptation, hoping that sociologists may take the hint and in their turn refrain from publicly discussing matters beyond their grasp. Rather, I should like to turn briefly to a very positive human aspect of the nuclear engineering community, viz., that of international cooperative programs. Cynics may scoff at this qualification and point out that such programs are merely the product of shrinking national budgets and rising costs. They may even be right. But I know for a fact that not a few of these programs, in addition to yielding significant technical and scientific results, have engendered lasting personal friendships spanning almost the entire globe.

I have already talked about the PISC program and shall now briefly mention a few others that I happen to know from close by. There is the OECD's Halden Project in Norway, originally emphasizing boiling water reactor technology but now turned into a breeding-ground of interesting and valuable ideas on control room display systems. The fracture mechanics round-robin program, again under the auspices of OECD (CSNI), aiming at an international comparison of elasto-plastic fracture mechanics computational results on standard crack geometries. And the International Cyclic Crack Growth Group, exchanging test materials for and information on experimental fatigue crack propagation under the joint auspices of EPRI and the USNRC. The willingness to exchange information in all these and many other international R & D programs should serve as a stimulating example to other branches of the engineering profession.

An activity also predicated upon the willingness of nuclear energy organizations to supply and exchange essential information is the formation of nationwide and possible international data banks on off-normal events in nuclear power plants, such as *e.g.* the USNRC-operated Licensee Event Reports (LERs). The central storage of data on the types and frequencies of failure of different kinds of equipment and complete systems, accessible for interactive interrogation by all participants, should make a significant contribution to the safety of nuclear installations. Successful operation will of course require experienced professional staff capable of solving the difficult classification problems inherent in this kind of work. Given such successful operation, however, these nuclear energy data banks could and should serve as an example for other high-risk industrial activities, to the benefit of society as a whole as well as of individual participants.

In the overview given so far I have hardly mentioned the effect of spin-off from nuclear energy on the generation of new jobs and businesses. I have refrained from doing so in order not to waste any more of your time than strictly necessary, considering this effect to be self-evident in all cases. Whether in the development of sophisticated NDE instruments, of software for data banks, or of new or adapted manufacturing processes, the emanation of sophisticated technology should continue to provide jobs and stimulate business requiring high levels of training, *i.e.* those jobs and businesses upon which the highly developed economies should concentrate for the benefit of the developing countries as well as of their own.

References

1. D.G.H. Latzko: Reliable heavy duty pressure components: The next 20 years. Mech. Engng. Publications Ltd., London (1980).
2. T. Tahara *et al.*: A feasibility study of manufacturing ultra-large pressure vessels for coal conversion application. Fourth Int. Conf. Pressure Vessel Technol., paper 66.
3. S. Sawada *et al.*: Application of narrow-gap GMA welding process to nuclear reactor pressure vessels. *ibid.*, paper 76.
4. M. Kawahara *et al.*: Advantages of narrow-gap GMA welding process. *ibid.*, contribution to discussion.
5. K. Fujioka: Strip overlay process and its greater productivity. *ibid.*, contribution to discussion.
6. ASME Boiler & Pressure Vessel Code, Section XI: Rules for inservice inspection of nuclear power plant components: Appendix A: Evaluation of flaw indications.

7. J. H. Vugts and R. K. Kinra: Probabilistic fatigue analysis of fixed offshore structures. *J. Petroleum Technol.*, **30** (1978), 4, pp. 657–67.
8. T. Belytschko and U. Schumann: Fluid-structure interactions in light water reactor systems. *Nucl. Engng & Des.*, **60** (1980), 2, pp. 173–95.
9. E. C. Rodabaugh: Review of data relevant to the design of tubular joints for use in fixed offshore platforms. WRC Bulletin 256 (1980).
10. J. Schroeder and P. Tugcu: Plastic stability of pipes and Tees exposed to external couples. WRC Bulletin 238 (1978).
11. American Petroleum Institute. Recommended practice for planning, designing and constructing fixed offshore platforms. API RP2A, 7th edition (1976).
12. Human factors in the nuclear control room. EPRI-NP-309 (1976). (Summary by J. L. Seminara *et al.*: in *Nucl. Safety,* **18** (1977), no. 6, pp. 774–790.
13. Report of the President's Commission on the accident at Three Mile Island. The need for change: the legacy of TMI. The Commission, 2100 M. Street NW, Washington, D.C. (1979).
14. E. W. Hagen: The human factor: Key factor in nuclear safety. *Nucl. Safety,* **21** (1980), 4, pp. 480–5.

Soft Energy vs Nuclear Energy

Yoshio ANDO

Abstract

During the early 1960s, a plentiful, inexpensive supply of petroleum enabled Japanese industry to progress rapidly; however, almost all of this petroleum was imported. Even after the first oil crisis of 1973, the recent annual energy consumption of Japan is calculated to be about 360 million tons in terms of petroleum, and actual petroleum forms 73 % of total energy. It is necessary for Japan to reduce reliance on petroleum and to diversify energy resources. The use of other fossil fuels, such as coal, LNG and LPG, and hydraulic energy, is considered as an established alternative.

In this presentation, the author deals with new energy, namely soft energy and nuclear energy, and discusses their characteristics and problems. The following kinds of energy are dealt with:
a) Solar energy,
b) Geothermal energy,
c) Ocean energy (tidal, thermal, wave),
d) Wind energy,
e) Biomass energy,
f) Hydrogen,
g) Nuclear (thermal, fast, fusion).
To solve the energy problem in future, assiduous efforts should be made to develop new energy systems. Among them, the most promising alternative energy is nuclear energy, and various kinds of thermal reactor systems have been developed for practical application. As a solution to the long-term future energy problem, research on and development of fast breeder reactors and fusion reactors are going on.

1. Introduction

Since the oil crisis in 1973, energy saving has become a worldwide catch

Department of Nuclear Engineering, University of Tokyo, Tokyo, Japan.

phrase, but the total demand for energy is still increasing because of factors such as population growth, elevation of living standards, and growth of industry.

In terms of petroleum, about 6,000 million tons of energy were consumed in 1975 in the world. In the U.S.A. and Japan, 1,780 and 360 million tons of energy, which form 30% and 6% of world total energy, respectively, were consumed as shown in Table 1.[1] In comparison between the two countries, Japan is distinguished by a very high import energy ratio and petroleum ratio, and more than half of its energy is consumed for production. World opinion requests a reduced petroleum ratio for Japan and reduced energy consumption per head for the U.S.A.

Table 1.   Energy consumption in the world, the U.S.A. and Japan.

Region	Total energy (million ton in terms of petroleum)	Percentage in respective total energy		
		Imported	Petroleum	Production
World	6,000	—	—	—
U.S.A.	1,780	18%	43%	32%
Japan	360	88%	73%	57%

Petroleum still remains the most important energy source, but it is no longer possible to rely on it as we have done in the past. Therefore, various alternative energy sources are being considered; some of them, such as hydraulic, coal, LNG, and LPG, are established energy sources despite their demerits. Reserving discussion on those established alternatives, in this presentation so-called new energy sources such as soft energy and nuclear energy will be dealt with and their characteristics and inherent problems will be discussed.[2,3]

2. Solar Energy

The sun is a gigantic nuclear fusion reactor 150 million km distant from the earth. The mass is changing into energy and the mass defect amounts to 5 million tons per second. Then at the outer atmospheric boundary of the earth, solar energy comes down continuously at the rate of 1,200 kcal/m²/hr—that is, 1.4 kW/m². On the ground at Japan's latitude, it is estimated at 0.24 kW/m² averaging day and night and considering attenuation by weather.

Solar energy is so vast that only 20 minutes of solar energy intercepted by the earth equals the earth's total annual consumption of energy; however, the energy density is low as mentioned above. All mankind and living things have enjoyed the benefits of solar energy from time immemorial.

By adding some simple equipment, the use of solar energy has been expanded. Cultivation of vegetables is possible even in winter in vinyl houses; hot water can be supplied by solar heat collectors, and air conditioning is possible in a solar house.

Electric power generation by solar energy can be done in the following two ways:

2.1   Photovoltaic Generation System

In this system, an electric current flow between electrodes occurs when sunlight strikes the combined P- and N-type semiconductors in a solar cell. Figure 1 shows a scientific satellite developed by the University of Tokyo; its surface is covered by solar cells which generate electric power for the sounding instruments and the telemeter. Because solar cells are of low efficiency, costly, and not suitable for mass production, they are used only for specialized applications. Development to counter these demerits is essential for wider applications in the future.

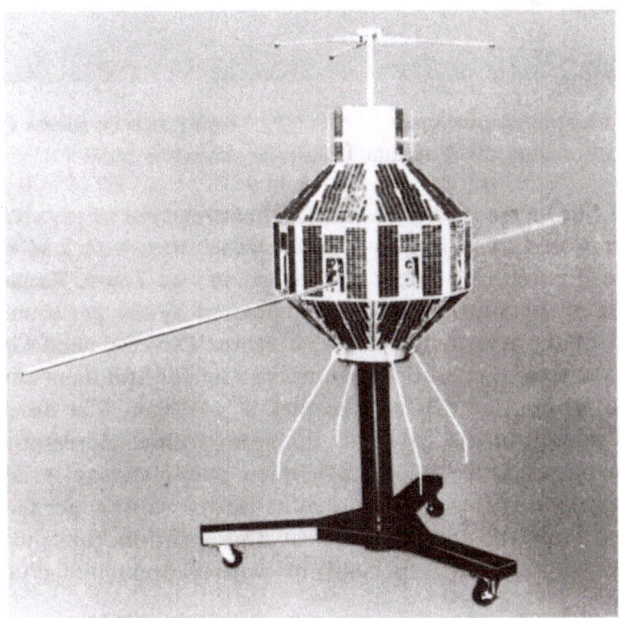

Fig. 1.   Scientific satellite developed by the University of Tokyo.

2.2   Solar Thermal Power Generation System

This system converts the radiant energy of the sun to thermal energy, which in turn produces steam of above 300 °C to drive the turbines and

(a) Central receiver type   (b) Parabolic mirror type
Fig. 2.   Solar thermal power generation plant.

generators. For heat collection, a central receiver-type or parabolic mirror-type system is used. A solar power generation system with 1 MWe capacity as shown in Figure 2 is under construction at Nio Town, Kagawa Prefecture, as part of the Sunshine Project sponsored by the government; it will be experimentally operated by Electric Power Development Co., Ltd.

Small-scale power generation can utilize roofs of buildings but for large-scale power generation, space becomes a problem. The necessary area for power generation of 24×10^6 kWh/day (which corresponds to the electric energy generated at a 1,000 MWe power station in 24 hours) is shown in Table 2. It is clear that a solar power station needs more than one hundred times the area of a nuclear power station; thus the land problem will be very serious, especially in densely populated countries like Japan.

Table 2.   Necessary area for power generation of 24×10^6 kWh/day.

Nuclear Power Station	0.34 km² *
Solar Power Station	41.7 km² **

* TEPCO Fukushima-II; 4,400 MW, 1.5 km².
** 0.24 kW/m², $\eta = 10\%$.

3. Geothermal Energy

Since Japan is a volcanic country, it has abundant potential energy reserves of geothermal energy that are estimated as one tera (10^{12}) watt in terms of power generation output, which is equal to about four times the present total energy of the country. However, under present conditions, a few domestic geothermal power stations totalling 200 MWe output are in operation, and the largest one is only 50 MWe capacity. With regard to generating facilities, Japanese units are widely used in the U.S.A., the Philippines, and so on, but most of the suitable sites for geothermal power stations in Japan are in national park areas where environmental restrictions are more strict than other places. There always exists a risk when boring is not successful in getting the expected result.

The power generation system today is limited to the use of natural steam, but for the purpose of more effective use of geothermal energy, the development of a binary cycle power generation system using a low-enthalpy fluid such as flon gas or isobutane as the carrier of energy, or a combined cycle system combining the former with hot water with the natural steam system as shown in Figure 3, is expected. Besides these, techniques for removing arsenic, hydrogen sulfide, and scales from geothermal brine to protect plant facilities are being developed. Overall feasibility study on volcanic and hot dry rock power generation is also in progress.

In the Japanese program, the total output of geothermal power stations will reach 1,000 MWe in 1990, which corresponds to only one unit of a nuclear power plant, so too much cannot be expected from geothermal energy.

4. Ocean Energy

Japan is surrounded by the sea, where various kinds of perpetual energy are reserved.

4.1 Tidal Energy

There are about two tide cycles every day, and the difference between the rise and fall of tides is related to the position of the moon and the geometrical shape of the sea near the coast.

In France, a tidal power station was actually constructed by damming up the mouth of a river where the tide difference reaches 9m. Reversible Kaplan water turbines which are connected to generators are located at the bottom of the dam as shown in Figure 4, and operated at high and low tide.

A feasibility study on tidal energy was performed in Japan on the Ariake

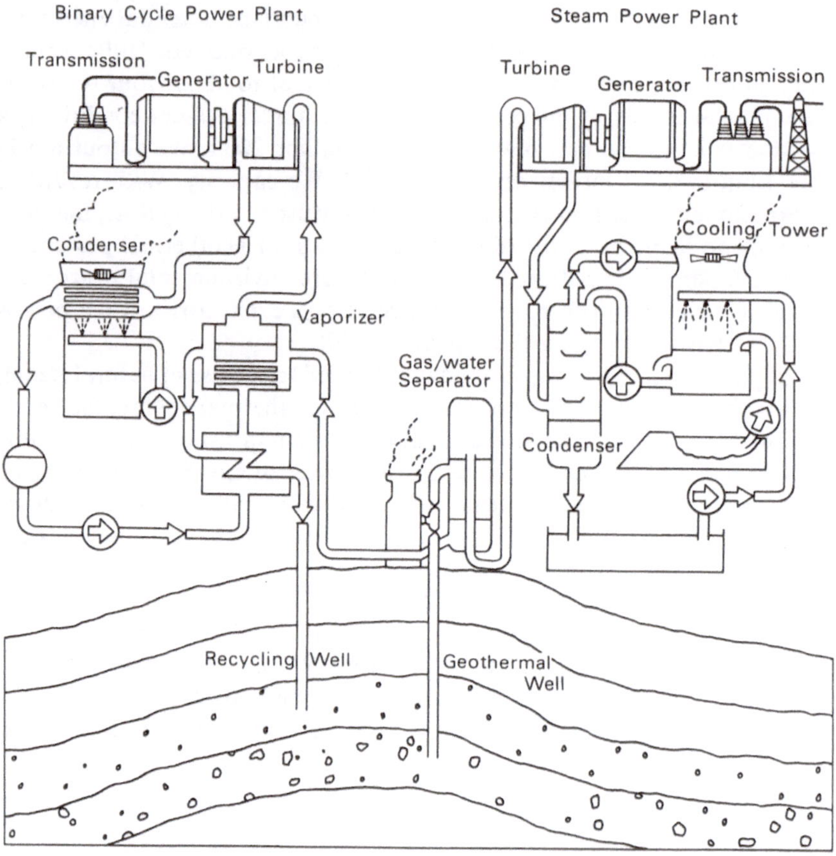

Binary Cycle Power Plant

Steam Power Plant

Transmission
Generator   Turbine

Turbine
Generator   Transmission

Condenser

Cooling Tower

Vaporizer

Gas/water
Separator

Condenser

Recycling Well

Geothermal
Well

Fig. 3.   Combined cycle geothermal power plant.

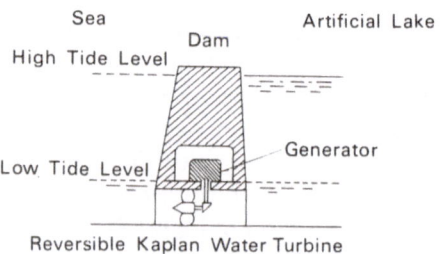

Sea

Artificial Lake

Dam
High Tide Level

Generator

Low Tide Level

Reversible Kaplan Water Turbine

Fig. 4.   Tidal power generation system.

Sea in Kyushu, where tidal difference is expected to be 5m. The results were negative, showing only 10% availability and about 100 times electricity unit price. Another demerit of this energy is the phase discrepancy—that is, the tidal phase coincides with the moon while people's life coincides with the sun.

4.2   Ocean Thermal Energy Conversion (OTEC)

The temperature of the zones near the surface is warmer than the deeper zones of the sea. The thermal energy represented by this difference is used to generate electric power. The principle is as follows: When a closed vessel with warm water is connected to a cold vacuum chamber, steam vaporation is initiated. This phenomenon is more clear when low-enthalpy fluids are used in place of sea water, and it is calculated that a temperature difference of 20°C is well balanced to generate power, taking off the energy necessary to drive sea water and vacuum pumps. The problems are the difficulty of finding a location where a great enough temperature difference can be derived and the environmental effect of circulating a vast amount of water vertically.

4.3   Wave Energy

Dynamic wave energy can be used to generate power. A few years ago, R & D work was started by the Science and Technology Agency at Tsuruoka on the Japan Sea. The principle is shown in Figure 5: the air in a bottomless chamber is compressed by a rising wave and then drives an air-turbine which is driven in the same direction by the air flow when the wave is descending. This R & D work qualitatively verified the possibility of power generation by waves; however, there are some problems, such as the cost of a facility and wave availability, which depend on the weather.

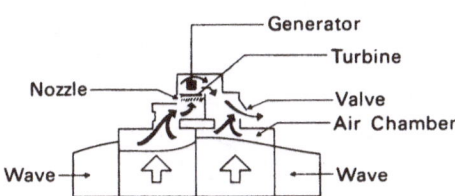

Fig. 5.   Wave force power generation system.

5. Wind Energy

Wind energy on a moderately large scale has been used by sailing ships for hundreds of years when none of the other kinds of energy found a way

to be utilized. Sailing ships commanded by Columbus succeeded in cross-
ing the Atlantic, and China clippers or Cutty Sark were the leading trade
cargo boats in their age. After the First World War, sailing ships gave way
to steam ships and motor ships, because of the difference in speed. After
the oil crisis, a motor ship with computer-controlled sails was constructed
to save energy; however, it will be difficult to recover its old prosperity.

R & D projects on power generation by wind, such as the Science and
Technology Agency's Futopia project, are under way. In the TEPCO project in Miyakejima illustrated in Figure 6, a windmill of 29m dia. can
generate 100 kW power.

The problems in wind energy are the difficulty of large-scale power
generation and of securing a stable power supply.

6. Biomass Energy

In Brazil, squeezed sugar cane is used as raw material for methanol
which is mixed in gasoline. In Japan, there are very few materials which are
suitable to produce biomass energy, but urban trush can be used as an
energy source, even though on a small scale.

Fig. 6.   Wind force power generation system.

7. Hydrogen Energy

Hydrogen has various advantages. Its heating value amounts to 29,000
kcal/kg, which is about three times higher than that of petroleum. It is

obtained from water and produces only water without emitting pollutants after combustion. It can be easily stored and transported, usually in the form of liquefied hydrogen gas.

However, it should be noted that hydrogen is not primary energy such as fossil fuel, natural energy, and nuclear energy, but is secondary energy. The situation of hydrogen is similar to that of a pumping hydraulic power station which stores energy in the form of hydraulic potential energy pumped up by residual electricity, and generates electricity whenever it is needed. Therefore, the energy which is necessary to produce some amount of hydrogen is always larger than the energy discharged by the combustion of the same amount.

8. Nuclear Energy

The most promising among the various kinds of so-called new energy is nuclear energy. Japan started development of nuclear power generation in 1955 and completed its first commercial nuclear power station, Tokai-I (GCR, 166 MWe), in 1966. After that, there were continuous efforts to construct light water reactors—that is, BWRs and PWRs. At present 21 nuclear power units (14,952 MWe), which produce 13% of total electric energy, are in operation (as of October 1980). All of these reactors utilize fission of ^{235}U by thermal neutrons, and by such use of uranium, resources are comparable with fossil fuels in quantity.

Natural uranium consists of 0.7% fissionable ^{235}U and 99.3% ^{238}U, which latter is very unlikely to produce a fission reaction. By capturing neutrons, ^{238}U is converted to ^{239}Pu, which is fissionable material. Plutonium is utilized in fast breeder reactors and also in advanced thermal reactors. Research on- and development of these reactors are under way and it is expected that they will play a main role in the near future. The resources of raw material for plutonium—that is, ^{238}U—are more than 100 times those of ^{235}U in the same uranium ore, but more effective use of uranium enables the use of lower quality ore. Then uranium resources can be estimated as 1,000 times more than the case where only thermal reactors are used for power generation.

In the distant future, when fusion reactors are realized for power generation, fuels such as deuterium, tritium, and licium can be extracted from sea water, so the energy problem of mankind is solved almost eternally.

9. Conclusion

It is recognized that we can no longer rely on petroleum as in the past and that we should develop new alternative energy sources, in order to assure stability and prosperity for society in future. It seems attractive that most of the new alternative energy occur in nature and are free from cost, but many problems exist in using them.

In electric power generation, higher prices are estimated for various kinds of soft energy than for oil-burning power stations. On the contrary, electricity generated by nuclear power stations is roughly half the price of that generated by oil-burning power stations, even today. Moreover, it is expected that further development of nuclear technology in fast breeder reactors, fusion reactors, and so on will solve the energy problem in quantity and also in price aspects.

The development of various kinds of alternative energy sources is essential; however, in the development of so-called soft energy, specific troublesome problems must be taken into consideration. On the other hand, nuclear energy looks like the most significant reliable energy source. Thus, to solve future energy problems, the main effort should be concentrated in the nuclear field, where it is needless to say that safety is especially important.

References

1. T. Fujimura: World Energy and Nuclear Development (Sankei Publishing Co., 1980).
2. Y. Ando, vs. A. Robins: Nuclear power generation or soft energy. *The Ushio*, No. 253, 1980.
3. Y. Ando: Various alternative energy and nuclear energy. *The Thermal and Nuclear Power*, **31**, 292 (1981).

Discussion: Part VI

Dr. G. Yagawa[*]

I agree with your view indicated in your lecture that the superiority of manufacturing techniques in nuclear industries in Japan—for example, forging techniques, welding, and so on—is well recognized throughout the world. On the other hand, it seems to me that regarding the software in nuclear structural engineering, like code development, the Japanese contribution compared with those of other countries—especially of the U.S.A.—is not so notable. The ASME Code developed in the U.S.A. is employed in Japan in almost its original form to design and inspect nuclear components. I would like to have your comments on this point and on how Japan can, if needed, play a role in software development in nuclear engineering, including design and inspection codes of components, and also on to what extent nuclear code developments in European advanced countries have diverged from the ASME Code.

Dr. D. G. H. Latzko

In reply to the last part of Prof. Yagawa's question let me point out that the European countries follow a similar approach to nuclear plant design as described by him for Japan: they essentially follow ASME Sections III and XI for LWRs. With minor differences this applies to all countries of Western Europe, both large and small, notably such advanced countries as France, the Federal Republic of Germany, and lately the U.K.

To this statement I should like to add as my own opinion that such an apparent lack of originality might and perhaps even should be looked upon as an advantage from the industrial standpoint. To take an example: had Germany adopted significantly different design rules of its own, the huge JSW forging shown on one of Prof. Ando's slides might never have been exported to Germany. This would in my opinion have implied a net loss, in that it would have deprived one country of the unique skills available in another country.

Contributions to what Prof. Yagawa referred to as the software part of nuclear structural engineering should therefore in my view concentrate on areas not yet fully and unequivocally dealt with in the ASME Code. Perhaps the most important example in this respect is the set of design rules given in the various consecutive Code Cases for structures operating in the creep range. Here I believe that Japan, like France, is in the process of making significant and much-needed contributions of its own. Other, more limited but not necessarily less important, areas for such complementary contributions are the assessment of residual stresses and the application of elasto-plastic fracture mechanics.

[*] Associate Professor, Department of Nuclear Engineering, University of Tokyo, Tokyo, Japan.

Part VII
Conclusion

Panel Discussion: The Role of Nuclear Engineering Research and Education for Energy Futures

Dr. K. OSHIMA

We now have come to the last part of the symposium. The topic of the panel discussion is education and research in nuclear engineering in the universities and the role they can play in energy futures.

In the past two days, lectures have been given in six sessions of this international symposium, commemorating the 20th anniversary of the Department of Nuclear Engineering, the University of Tokyo. These lectures and the discussions which followed them can serve as important guidelines for the discussions of the panel.

The professors invited from abroad to this international symposium all have a close and long relationship with our university. In this panel discussion, we would like especially to take up the role of education and research in our department at the University of Tokyo for the future. We have selected three sub-themes.

As the first subtopic, we would like to discuss the current status of nuclear engineering education at universities and what should be the direction that we should move—in other words, the problems we face in the current situation and the future perspectives of nuclear engineering education.

The second subtopic is basic research in nuclear engineering: what should be the direction of the future, and what are the current problems.

The third subtopic is safety research at universities. Universities have to maintain a neutral position in the society. In this respect, universities can play an important role in public safety, and safety research can be a very important area of concern at universities.

After a brief introduction to these three topics, we would like to have a general and overall discussion. We have three panelists, Profs. Sekiguchi, Takahashi, and Kondo, to present the current situation at the University of Tokyo, as an introduction to the discussions which follow. After that, I would like to invite panelists from abroad to make comments. Also, I would like to open the floor to participation by the audience. Now, without further ado, I would first like to call upon Prof. Sekiguchi to speak about the problem of education in the area of nuclear engineering.

Dr. A. SEKIGUCHI

The Japanese university education system has undergone tremendous change since the War. Before the War, universities in Japan were modeled after European universities; after it, American universities became the model. After three years each of junior high school and high school, the undergraduate university course is four years. At our university, the first two years are devoted to general education and the last two years to specialized study. The problem in each specialized discipline, therefore, is the course of study for that two-year period.

A problem of nuclear engineering is that the technological field which should be learned by students for such limited period is so broad that it ranges from nuclear reactor technology to biological application of radiation, for instance. The nuclear reactor technology itself includes the design, construction, and operation of fission and fusion reactors as energy-generating systems.

Our classroom lecture subjects include plasma science, nuclear reactor physics, nuclear radiation instrumentation, structural engineering, materials science, thermo-hydraulics, nuclear chemistry, chemical engineering, and nuclear fuel cycle engineering. These are interconnected to form a consistent integrated system. Undergraduate students are recommended to learn most of these for their two-year period.

As for the graduate course, while majority of students have finished the undergraduate course of nuclear engineering, some come from other departments of the university—electrical engineering, mechanical engineering, applied physics and chemistry, or materials science—whose courses of study overlap somewhat with that of the nuclear engineering department. So we have two orientations: one is to give students a basic grounding in the field, and the other is highly specialized training in a few limited areas. When we ask graduates of the nuclear engineering department, they say they do feel that their knowledge is more limited in scope than that of students in other engineering departments. On the other hand, it is important to keep up with the enormous degree of technological change taking place all the time in the nuclear engineering field.

In the past 20 years, about 520 students have graduated from the University of Tokyo nuclear engineering department; 250 obtained the master's degree, and 80 the doctorate in nuclear engineering. In Japan as a whole, every year bachelor's degrees in nuclear engineering are obtained by about 300 students, master's degrees by about 130, and doctorates by about 20.

At our university, about 40% of the bachelor's degree recipients go on to graduate study; about 40% find jobs with manufacturers, and 20% enter utilities, government agencies, or research institutions. Of the master's degree holders, about half go on to the doctoral course and the other half find jobs in the industry or in research organizations.

We recently devised a questionnaire for the graduates from our department, asking what they are doing in their current jobs. About 40% of them are involved in software design or planning; another 40% are working in the area of construction, manufacturing, and hardware design, and the remaining 20% are involved in basic research. A slightly different breakdown shows that about 40% work on reactor physics, reactor design, and safety analysis; and about 20% work on materials and fuel.

When nuclear engineering graduates go to work for a manufacturer, they typically specialize as follows: about 20%, the highest share, in plant equipment; the second highest share, in reactor core design; the third highest, in instrumentation and control; the fourth highest in safety; and the fifth in fuel problems.

Viewed from the industrial standpoint, our graduates are very much concerned with the management of projects; they are regarded as good coordinators or organizers. They are also heavily involved in research and development.

At the same time, a popular view among industry people is that our graduates are very well trained at software, but less well trained at hardware. Core design, safety analysis, and software applications are their strong areas; plant equipment design, electrical and mechanical engineering, nuclear fuel process, chemical and meteorological engineering aspects, and so on are their weaker areas. Therefore I believe we should place increased emphasis on these hardware-oriented areas in coming years.

Another point often made by private industry is that we sometimes tend to overspecialize the students in an area in which technology is changing at a rapid pace—for example, nuclear fusion. As a result, they are unable to find jobs exactly fitting their knowledge, and the companies which employ them are required to retrain them. From the industry's point of view, it would be better for us to concentrate on a solid basic foundation upon which the companies can then build in their on-the-job training programs. Japanese companies customarily give new employees one year of such training, rotating them to all parts of the company so they get a complete picture of the product or service.

I believe it is important to clearly identify the areas in which nuclear engineering education can make unique contributions: in the safety-related areas, for example, or in the study of phenomena which occur inside reactors as a result of the interaction between temperature, pressure, radiology, magnetic field, and so on. These areas are those in which nuclear engineering students should specialize.

It is also important to facilitate cooperation among universities, other research organizations, and industry. Joint research is essential: nuclear research projects are often gigantic in scale, and it is impossible for one organization to do everything. I personally feel very strongly the need for good communication between university and industry, and would like to appeal to both sides to work at establishing a good basis for communication and coordination.

Dr. K. Oshima

Thank you very much, Prof. Sekiguchi. I believe that you have given us a good summary on the current situation and the status of education.

Compared with the status of education in the Department of Nuclear Engineering at the University of Tokyo, explained by Prof. Sekiguchi, I believe that the situation may be different in foreign contries. I would like to ask the foreign participants to present their views if there are any similarities or differences.

Dr. M. Benedict

First, one has to do with the relative emphasis to be placed on undergraduate and graduate education of nuclear engineers. At MIT, we were very reluctant to introduce nuclear engineering into the undergraduate curriculum because we felt it was a specialized field, which was better taught after the students had learned one of the more traditional engineering disciplines. However, we found that both from the students' standpoint and from the standpoint of potential industrial suppliers, there was a real need for an education at the undergraduate level, and about three years ago we instituted an undergraduate program also.

The program has been moderately successful, but it still partially confirms our view that the main role of technical universities such as MIT is to train people at a more advanced level for nuclear engineering professional work, and that the preferred educational pattern is for the student first to learn a traditional discipline such as mechanical or chemical engineering, and to become a more specialized engineer at the graduate level. This is not to say, though, that the undergraduates, if they wish to leave the university to take a position, say, in an electric power company, cannot play an important role in the company. But they are not likely to make original contributions to nuclear engineering without more advanced training.

Dr. T. PIGFORD

Yes, our experience at the University of California at Berkeley with regard to undergraduates is very similar to that which Prof. Benedict described. We did, recently, very carefully try some undergraduate work. First, because of the same thing he mentioned, we organized the program so that the students in four years could satisfy the degree requirements in one of other fields like mechanical engineering and also nuclear engineering. That gave us considerable comfort. It's a hard program. We have such a program with almost every other department. The most popular ones are electrical and mechanical, each dual degree program with nuclear. They attract only best students and they are quite successful. We now have a single degree program also.

I want to add something about my view of the trends in education with regard to curriculum. In the United States, we have constantly asked: What is the purpose of a nuclear engineering department? It was easier to identify it as reactor physics because no one else taught that. But we soon realized that the nuclear engineer must learn about heat transfer and fluid mechanics, thermal dynamics, materials, some chemical engineering. But, of course, those are taught in other departments. We don't really mind, because, similar to the evolution of chemical engineering which, in the view of some people, started because of lack of attention to those problems from mechanical engineering, nuclear engineering also has the purpose of teaching these things.

What is its purpose? In my view, the nuclear engineer must know thoroughly the principles in those areas I mentioned. In particular, he must know the interrelationships, how to synthesize them, how to make them fit together. And I think that is usually his function in industry also.

Now, a problem is how you teach that. It's very easy to select specialists to teach reactor physics, fluid mechanics, and so forth. But fitting together, striving for that is a little harder. For years we have had a special effort on this. In fact I was privileged to participate in that sort of effort when I worked under Prof. Benedict at MIT, through rotating the teaching assignments; in those days we taught everything and we had to learn it. At Berkeley, we try consciously to do this still. Now it's dangerous; some professor from physics can do a bad job in materials. But if he's carefully motivated and he gets help, he will do a good job. It surely gives him the feeling of the other things, and when he teaches his regular course he can put into that course these other considerations. I think that is an extremely important thing to try. It takes a lot of time and a lot of effort. Also we bring in lecturers from industry, which I know is done in the Japanese universities.

I've had the privilege of visiting most of the Japanese universities' departments of nuclear engineering. I think they cover some of the fields that are lacking in the United States: materials. I think materials are so important, and there are some materials, the phenomena fundamentals of which are not being taught in other materials departments, need to be taught in nuclear engineering department and are so important to this synthesis. There's not enough of that in the United States.

I believe that, regardless of a textbook which is now several decades out of date in nuclear chemical engineering, there's not enough pure fuel cycle work in other universities. I find only one place in the country that carefully teaches structural engineering in a nuclear engineering department: that's MIT, and they do a fine job. And then there is waste management, and as of last spring there was no university in the United States except Berkeley which was trying to move into this field.

Dr. K. Oshima

Well, in Japan we do have the Tokyo Institute of Technology modelled after MIT, which only takes graduates and no undergraduate students but the University of Tokyo has undergraduate courses. So there are differences even within Japanese universities. Perhaps there are comments from the Japanese participants.

Dr. Y. Takahashi

I was very much impressed to hear Prof. Pigford's comment, in particular, that the nuclear engineering educational program should cover subjects from materials chemistry to reactor physics. We were wondering whether we should teach those two subjects to all students or whether we should divide students into two groups: one emphasizing reactor physics, the other group emphasizing materials or chemistry. We have come to the conclusion that it will be beneficial for all the students to take up both reactor physics and materials aspects, though it is a very hard program for them. I think this is current way of thinking.

Dr. T. Pigford

I'm inclined to agree with Prof. Takahashi, that it's most desirable for students trained primarily in physics to learn something about the other side of the discipline and students trained primarily in the chemical or materials aspect to learn something about the neutronics and physics aspects of reactors. Not all students are willing to do this unless forced to do so. Our policy has been not to force them but to urge them strongly, and I think it has the effect of obtaining, for the students who can benefit from learning the other disciplines, an opportunity to do so. But we do not require it of all students.

Dr. D. G. H. LATZKO

I have practically rather no contribution of my own to make, for two reasons. First of all, in the Netherlands, we have no bachelor's degree, so engineer's degree means a master's degree. That's one reason why I can't contribute. The second one is that, being a small country with no reactor development of its own, we have no use at all for a nuclear engineering department. However, we have a participant from West Germany, which is a large country and which has a sizeable nuclear development of its own, and I happen to know that they do not have a nuclear engineering department anywhere; they follow the course of training mechanical engineers, electrical engineers, and chemical engineers and then having them specialize on the job. So, my suggestion here would be to ask Dr. Jahns to make a contribution if he so wishes.

Dr. A. JAHNS (floor)

Professor Latzko is completely correct. That's what we do have in Germany. First, at universities and technical high schools, the education is for a master's degree, or a diploma degree in mechanical engineering, electrical engineering, and so on. And then we will have additional courses in nuclear physics and other different fields in the whole field of nuclear engineering where the students can specialize, normally in the second half of their educational term, to get the master's degree. This is the major pattern of education in Germany. On the other hand, we have a kind of bachelor's degree, not completely comparable to the American bachelor's degree. This is the so-called graduated engineer, just translated from the German. This is an education much shorter than the education at the universities, for normally two and a half to three years, and the requirements are not as high as the education requirements at the universities. And, of course, the major part of nuclear engineering education in Germany is, then, on the job wherever you go.

Dr. T. PIGFORD

I understand the question we are responding to is to what extent should there be requirements in specified areas, like, for example, reactor physics, chemistry, materials, and so forth. Berkeley was the home of the student uprising that began on the Sproul Hall steps in 1968, and in response to that we tried having completely unstructured curricula for a while. It appealed greatly to students, philosophically. We even have one campus that specialized in that, and that campus is now having great trouble attracting good students. We have learned since then that we have an obligation to tell our students, if they want to come into nuclear engineering, what is a professional nuclear engineer; that's why they come to us. And one way is telling them what they must know. You can translate this in terms of some kind of requirements: take these courses and pass them. In my view, we owe it to students to very carefully look at what they must know in order to be a nuclear engineer from the University of Tokyo or from the University of California, and it may be different. It is good when it is different. But each university, I think, has that responsibility. In my view, it is very important to implement that.

Dr. D. G. H. Latzko

We've heard about the German case from Dr. Jahns and it would be equally very interesting if we could hear Monsieur Petit's opinion about France, which also has a very sizeable nuclear engineering effort. In fact, I think they're close to being world leaders. To my knowledge, there is no nuclear engineering degree, but there is excellent postgraduate training at Saclay which then takes the character of a nuclear engineering postgraduate course. And my knowledge of French nuclear engineering is sufficient to say that it is second to none. It would be interesting to have Mr. Petit's comment on the fact that they, like Germany and Holland, do not have a nuclear engineering degree and yet have good nuclear engineers.

Dr. J. F. Petit (floor)

The French situation is exactly as Prof. Latzko said. The only exception is that, only some three or four years ago, one technical university was created in France, the Compiegne University, and this university now has its own department of nuclear engineering. By the way, the President of this university is a famous nuclear man, Monsieur Guy Danielou, who played a great role in the past in fast reactors in France.

Dr. K. Oshima

Since we have been discussing synthesis at our department, Prof. Kondo may like to make some comments.

Dr. S. Kondo

My comments are related to what Prof. Pigford mentioned: the principles and theories that the department of nuclear engineering has to offer to students. Needless to say, jobs in nuclear industry are not limited to or cannot be fulfilled by the graduates of nuclear engineering departments alone. So the question is what the graduates from nuclear engineering departments can offer to the whole endeavor of nuclear energy development. We have discussed this question within our department since its establishment. And we have decided to intensify the education of synthetic aspects of nuclear engineering so as to give the student an understanding of the interrelations among the various disciplines.

As a part of this change of weight, we have started a one-semester design project course. In this program a team of about ten students is given a design problem with a few items of specification. They analyze the problem jointly and organize several groups for specific design tasks under one student whose role is to manage the project. I think in our program they can really come to grips with what it is to be a nuclear engineer. The students are usually very enthusiastic and spend even their own time to try to complete the project, and they come up with very good output.

Of course this kind of synthesis activity alone does not do the job, and I am not saying that this project, which may be called a kind of on-the-job training within the university, reflects the major objectives of education. What I would like to stress is that this project provides the student a very good chance to understand not only one aspect of practical engineering but also the importance of the so-called interrelations Prof. Pigford has mentioned, through the understanding of the principles in various well-established disciplines.

Dr. K. OSHIMA

Are there any comments from other participants? Perhaps we can touch upon the relationship between universities and industry. What are the things that industries are expecting of us and of the graduates who go into industry? Of course they will receive on-the-job training. So, can we, the universities, leave it to industry? Or do we have to prepare students so that they can receive better on-the-job training?   Can anybody comment on that?

Dr. T. PIGFORD

If I understand your point, then it's an important point: what is the function of the on-the-job training in industry? I think that it is to make engineers out of them. I don't think I can pretend that the graduates from our departments are professional engineers. They have been taught a few things, and we try to teach it to them well. But engineering is the practice; it is the practice where you finally have the synthesis. And the people who finally must teach you are the people who are doing it. Of course, some people learn it in a self-taught way, and we can see some good results from that and some disastrous results. So the industry has a major role. I happened to talk yesterday about the danger of new graduates going directly to work for a government agency like a regulatory commission if they don't have that kind of experience. It's necessary; maybe the government agency can or could provide it. So, I think that we must recognize that in education we can only do part of the job of making them professional engineers, and it's very important that the rest be done.

Dr. A SEKIGUCHI

I have listened with great interest to the discussion of whether or not students who immediately join a government agency such as a regulatory authority are capable of doing their jobs on their own. In our department, actually, two or three of the undergraduates join government agencies per year. While these government agencies, such as regulatory authorities, do give some sort of on-the-job training, I think this is a problem that we should give more attention to.

Dr. T. PIGFORD

I do agree with Prof. Sekiguchi. Since this is a rather sensitive area perhaps, let me try to make it pretty clear. I don't think there's necessarily anything wrong with a new graduate joining a regulatory agency or some other government agencies, provided that agency does for him what he would get if he also joined industry—provides for him the integration, the synthesis, and the experience so that he can be a professional practicing engineer.

Dr. D. G. H. Latzko

It just occurred to me that I might make a contribution myself. We've been discussing how to expose the graduate to practical engineering. Now, what I'm going to say may sound provocative, and it's not my purpose to be provocative. But in many European countries like my own, if you become a professor, you become so after about twenty years of industrial practice. That, in some cases, may be an advantage because it's taught me humility. It's taught me actually how difficult it is to operate a power plant, to commission a steam turbine plant, and therefore, quite seriously, I'm wondering why the European practice of having at least a sizeable number of full professors coming from industry might not be considered elsewhere. Alternatively, one could use agent professors, which is the way I used to work for a couple of years. I'm not saying that this is the only way, but I'm just mentioning that perhaps it's a way of bringing practice into the university.

Dr. T. Pigford

I think a very important thing is that faculties in the United States do not have enough practical experience in the field they are teaching. I think it has gotten much worse since World War II for various reasons, but that's the fact. There are certain unfortunate administrative reasons why professors are not brought in very much at the higher levels, and adjunct professors are useful but they are not the core of the teaching. So we must find some ways of dealing with it; either change the system or find other solutions. I think professors' consulting is extremely important; they bring back to their work knowledge of up-to-date problems, and that is one approach.

Dr. K. Oshima

Then let us proceed to the next subject. I'd like to call upon Prof. Takahashi to speak about research.

Dr. Y. Takahashi

I think education is basically nurturing young people who are expected to play important roles in the future. That means students have to cope with present problems, and at the same time they have to be taught to be able to cope with future problems.

We at the University of Tokyo have developed an undergraduate curriculum through trial and error. Nuclear engineering proper is not the only subject. We try and offer a very broad range of subjects so that the students may begin to understand particular phenomena from the basis of a broad background.

At the graduate school, we do not have clearly defined requirements. At the graduate level, students have an advisor who tries to bring out their best potential. Thus each student's curriculum is designed to draw out the best in him.

We have that sort of unstructured system at the graduate school level. This should be conductive to an innovative type of basic research, and I think the curriculum we have now is very conducive. Each researcher can freely select and choose a topic, and he is given a broad time frame to pursue that topic. Along with this, he has his own laboratory with certain characteristic features.

Nuclear engineering is task-oriented; this means that it is important for researchers to collaborate in a task-oriented project. But I think it is also good for a postgraduate to lose himself in one particular topic. It contributes to his development as an engineer.

Our nuclear engineering department covers a broad spectrum from chemistry to physics, and we have different researchers covering different fields. Each has his own idea, his own contribution, his own unique philosophy. There is much benefit for the department from those different positions. At the same time, to develop nuclear engineering science it is most desirable if all these ideas can be synthesized. So that is the framework, within which we try to do our best.

Now, what are some of the practical things one can do in basic research at the university? When it comes to practice and when it comes to implementation, you can't always live up to your ideals. It's more difficult. You can't own a power reactor. You're constrained in that way. You're limited to basic research and can't really go beyond that in many cases.

Some of our research efforts are related to the Nuclear Engineering Research Laboratory at Tokai Mura. There, we have been able to put some of our efforts into practice. One is Yayoi, which is the fast neutron source reactor. Quite a number of people contributed to the design work, and from that experience it was a contribution to engineering. At the present, it is used for basic research, not only by the University of Tokyo but also by other researchers and professors from other universities. This reactor has been created by many people, and it would be very effective if we had a system where the results and the data obtained from all the experiences and experiments could be fed back to all those interested.

Another effort is the electron accelerator which has been in operation for the last three years or so at Tokai Mura. This is not the exclusive effort of the University of Tokyo but really a joint collaboration with other institutes. The electronic pulse we can obtain from the machine is on the order of picoseconds. Specifically, pulse operation is achieved by using the picosecond radiation initial process. Without a very unique engineering effort, such a high level of performance would never have been reached.

Lastly, we have a tremendous commitment to the CTR Blanket Engineering Research Facility and we are pursuing research there. Chemists and physicists are contributing their past experiences, and that's again a collaborative sort of effort. I hope that the result of that effort will be fed back to benefit our basic research.

"Basic research" carries with it quite a number of different nuances. To some, it is a romantic academic image, while to others it is a useless and redundant sort of activity. In this respect, I think, we do well to be humble and listen to what the industry has to say, or what other people have to say; if we lend an ear to their constructive criticism, we can modify and revise our direction. That is going to benefit us.

Dr. M. BENEDICT

I'm inclined to agree that research has an important place to play in the education of engineers, that's the general philosophy at MIT and certainly is the philosophy of my department. But, for the reasons that Prof. Takahashi has just mentioned, it is difficult to provide realistic research facilities at the university. They're expensive, and in some instances they require very special precautions to take care of the nuclear hazards that are associated with them. So, at MIT, we've had to compromise on this.

One solution we've had for providing realistic engineering research opportunities for the students has been to undertake a cooperative program with the Oak Ridge facilities of the Department of Energy, mostly in the Oak Ridge National Laboratory, where the chemical engineering department at MIT for many years has operated an engineering practice school of which Prof. Pigford was once the director. We've made it possible for graduate students in nuclear engineering to enroll in this practice school also. And they obtain really first-rate on-the-job engineering research-type experience, working on the real live problems of the Department of Energy's complex at Oak Ridge, Oak Ridge National Laboratory Gaseous Diffusion Plant, and the various engineering facilities at the K–25 Plant. This takes them away from the campus, but at the same time it gives them a real-life-type experience that they couldn't possibly get in a university environment.

But, in addition to this, MIT has at least two facilities on campus where the same kind of real-life engineering experience has been available to the students. One of them is our own MIT reactor, which has gone through two design and construction phases and provided almost as good practical experiences for the students as working for an engineering design company would have been. In addition, they had the handsome opportunities of bringing the reactor to power and carrying it through its break-in operations and seeing it going into service. However, this isn't something you can do for every year's class; it was a unique opportunity for the students who happened to be there at the time the reactor was built and then in 1978, when it was reconstructed. The routine operation of the reactor does provide educational experience of another character for those students who would like to participate in the day-to-day operation of the facility. It's less educational if you wish. But at the same time it's good training for students who may wish to take similar positions in industry later.

The other MIT facility which students recently have benefited from working with is the Alcator accelerator, the Alcator fusion demonstration unit, if you like. This is MIT's contribution to the U.S. fusion program, where we provide a higher flux of particles but at a lower energy than is necessary for a self-sustaining fusion reaction. But at the same time it provides a real experience in the design and operation of a magnetic confinement system with all of the problems that a fusion community is going to have to solve before we have a practical fusion device. This is an ongoing program, and for students in our own fusion option it's a very fine way of obtaining handsome research experience. So I feel that this kind of an opportunity to participate in real-life engineering programs is a particularly important aspect of a nuclear engineering education, which I wish all of our students could make use of.

Dr. D. G. H. LATZKO

Prior to making any comment, I have a basic question, because if I make a comment it would be on both points two and three. Why have we only been talking about basic research at the universities? What is the reason for limiting university research to basic?

Dr. Y. TAKAHASHI

Well, I don't think we have tried to limit ourselves to basic research. But as far as the nuclear engineering department is concerned, as was mentioned previously, it is very hard to develop the sorts of technologies which are really applicable as they are for actual plants. Therefore, our major efforts have been devoted to basic research so far. At the same time, basic research is the most suitable area for the universities while very organized applied research can be done elsewhere. Therefore, I think this is the most important contribution universities can make and other research institutions cannot. For these two reasons, basic research at the universities is very important. It depends certainly upon the definition of basic research. When we talk about basic research, it doesn't mean scientific research but basic engineering research.

Dr. D. G. H. LATZKO

I think that I agree with Prof. Takahashi's second reason. As to the first problem, it is agreed, but I'd like to emphasize that significant practically-oriented research has been carried out, is being carried out, and hopefully will be carried out at universities.

My own experience throughout many years in dealing with just one little part of the University of Tokyo is with Prof. Ando's laboratory, and I'm not trying to single out a laboratory; I'm just saying that's my experience. A lot of the work which has been going there on is very, very much of direct applicability to engineering.

I think that's the third point, a safety. Many people, when thinking of the safety of reactors, are thinking of LOCA (loss of coolant accidents) and excursions. But I think about safety as a former power station superintendent. I think what I was most interested in was that the reactor runs with a load factor of about 80%. If that is what you are after, then some of the work which is being done in Prof. Ando's laboratory, and maybe in other laboratories as well—namely, finding out how far you can permit a crack to proceed before you have to shut down the plant—can be done very well at the university. It is being done at the university, and is also essential for giving the students a feeling of practical engineering because it teaches them what it is really about in practice.

Dr. T. PIGFORD

With regard to basic research, it is a problem in the United States. At the universities, we worry about obtaining money for research. Now, some of the federal agencies which also give out money for research have a charter mission to support only basic research, and to them this is unapplied research. It is very difficult for our engineering departments, particularly nuclear engineering departments, to have access to those funds. Why nuclear engineering? Because in the old days the Atomic Energy Commission took over the job of supporting research. It had a research division, and that division was told to support basic research and not applied research. In my view, we shouldn't get hung up on whether it's basic or not, but it is hung up in the United States.

Our research should be, frankly, applied. That's engineering. It should have the purpose of understanding phenomena. I think in that sense it differs from what I see being done by JAERI (Japan Atomic Energy Research Institute) and PNC (Power Reactor and Nuclear Fuel Corp.) at Tokai Mura. At a university, however, we have the privilege of spending more time and the responsibility of penetrating deeper for understanding. So, I think that's what distinguishes the kind of research we should be doing.

Professor Takahashi asked, "What are the views of industry?" I can tell you the frequent view of industry in the United States, which I respect very much. They would like to hire our graduates who have taken the courses, but they don't care about that extra two or three years on a research problem, speaking about graduate service. On the other hand, I think there we must resist them because it is part of the education, and we know it is, and it is a lifetime experience, as Prof. Benedict said. And so we should insist that it be part of our graduate program.

I think, besides getting money, a most important question is what to work on. Now, I think, in American universities, it's very easy to work on things that are not relevant, sometimes under this umbrella of basic research, because that had a great zeal after World War II. We do a disservice to our students. I think we should not be embarrassed if it's a problem that is practical or theoretical, or a year term or longer term. It should be relevant in some way, and we should work on it for understanding.

I want to mention a new technique which we've tried at Berkeley. We try to be able to move into new fields when they are needed. We worry greatly about relevance. And we have seen the great emphasis in the United States upon solving the technical problems of radioactive waste management. To some of us who came from chemical engineering, what are the missing things? What is the way of analyzing the long-term performance of a geologic repository? Too many decisions might be based upon plausibilities or speculations. And so we saw that this is very similar to problems in chemical engineering: chromatography. But more complicated. That is interesting.

To get started quickly, we were fortunate. One year we had Mr. Maeda from JAERI, who came to spend a year helping us get started and was a great benefit working with our students. The next year, Prof. Higashi from Kyoto University was a great benefit to us, and I hope he brought back some ideas to Japan. In fact, perhaps it worked very well because the next year, four people came, Mr. Masuda from PNC, Prof. Harada from Kyoto University, Mr. Iwamoto from Nikki, and Dr. Muraoka from JAERI. That team is now famous in the United States; it is called the SAMURAI team, and our students benefited very much, because, with the help of these people who were there also to learn something about a new field, we were able to establish a program quickly and very effectively.

Dr. A. JAHNS (floor)

In my opinion, the participation of students in the research at the universities is a very important part. I'd like to describe the situation in Germany a little bit. Most of the universities which do have nuclear engineering departments are related to the nuclear research centers, like the University of Karlsruhe to the nuclear research center in Karlsruhe, like the University of Aachen to the center in Jülich, like Berlin to the Hahn-Meitner Institute.

I could give several examples of that. The students, when they are at the universities, either in the graduate or the postgraduate phase, normally do their diploma work and their experimental studies at the research center and participate in the research at the center. The directors of the institute mostly are members of different departments at the university. So there is a very direct and real connection of the students and participation in their research done at the centers. The distinction between the research at the university and at the center is pretty hard to make. I have no numbers on this; it is a matter of funding and is hard to describe.

Dr. S. AN (floor)

Let me address myself to this issue of the research at universities. I believe there are two aspects: one is education and the other is research. I think these should be intertwined with each other. In other words, education at universities can only be realistic when research accompanies it. This is my first point.

Secondly, with regard to the issue of basic research, let me mention my personal experience. For example, fast breeder reactor development is being done at PNC in our country. A certain target has to be attained in a limited length of time in this sort of project. Some safety margin has to be allowed for actually designing and constructing the FBR. Along with these requirements, it is important to have basic research to further introduce various innovative elements to this. In that sense, the research being done at the universities is very important, and will lead the way to the successful development of the FBR project, because it really cannot be done by the practical engineers. This is the role to be played by universities. In the process, students or researchers at universities can get practical data and can be involved in more innovative experiments or research. I think this is the kind of basic research that Prof. Takahashi has been proposing, and, in that respect, what Prof. Takahashi was saying is in agreement with what Prof. Latzko is suggesting.

Another point I'd like to call to your attention is that in addition to making contributions to a specific project, we can also do research that is important on the engineering problems which are really basic and that cannot be done elsewhere. Especially in Japan, or perhaps in other countries as well, the role universities can play is to do basic research, i.e., to provide the basis or the foundation for further development of the technology of the future.

Dr. T. Pigford

I agree with Prof. An's points, and I think they are very important. I think there's one more thing to add. Many American universities have worried greatly about the seeming lack of facilities for our research on nuclear engineering. Now, when we initiated the program at the University of California at Berkeley, we worried about that, and so we made a large investment, and in addition to that investment there have been quite unique and remarkable facilities in adjacent laboratories which are used by our faculty and students.

But, looking back upon our experience during the last twenty years, I think some of the best work has been done in the facilities which were not those special ones I mentioned. Plain laboratories. Some good heat transfer experiments leading to understanding, some very nice materials experiments leading to understanding: Those required just a good laboratory and some money. I think excellent work can be done, and sometimes it presents a greater challenge to the professor to design such experiments. Sometimes the challenge yields better results. Last week I visited the Tokyo Institute of Technology and again visited Prof. Takashima's laboratory, seeing the beautiful experiments he was doing on absorption of iodine. Now, I've wanted these data for years on that kind of information. The big experiments at Oak Ridge did not lead to that understanding. There's another one; I've seen similar things in your laboratories in the Tokyo, Kyoto, and Kyushu universities.

Dr. K. Oshima

Thank you very much. Time is running short. I think we have to invite Prof. Kondo to address himself to the question of safety research.

Dr. S. Kondo

Nuclear safety is indeed one of the applied research problems we have just discussed. Needless to say, major safety research is conducted by the industry and government. Then what is the role of the university? Three points can be made. The first is that the university can contribute to industry by providing deeper insight into the phenomena they are concerned about, which I am sure is the basis for what will be done in the future to overcome any problems or to improve their present operations. The second point, which is related to the safety question, is the question of how to communicate with the public at large. We universities are recognized as the third party, and therefore it is necessary for us to provide objective judgment and offer it as an input to the safety discussion or debate. The third, but not least, point is that the university is free to discuss, design, and even do some experiments on novel ideas for safety enhancement. By doing so, the university can be one of the important sources of innovation in the development of safe technology.

Now for universities to perform these roles, what can we do specifically? Let me describe the problems that we are facing. The first problem is what specific areas should receive higher emphasis. Different people would have different ideas about the importance of each aspect. But we have to give priorities to the items that we can contribute to most effectively. Secondly there comes the question of scope or extent. To what extent can we make meaningful contributions to nuclear safety? It really depends on the capabilities and facilities that we can use. These two are hard but important questions we have to address ourselves to.

Thirdly, again, while not limited to the safety question alone, I think universities can provide courses for the people who are working in the industry. I am sure that to provide industry people with opportunities for retraining and re-education would contribute to sound growth in the nuclear industry. And last, particularly when we discuss safety, we have to strike a balance with international occurrences. For example, when the Three Mile Island accident occurred, it had a great impact on Japan's nuclear scene, and we are now reviewing our criteria for nuclear power plant siting partly because the U.S. changed its criteria rather drastically. This is just one example of how external events can influence the domestic nuclear scene. So a certain form of international coordination will be required. There are many formal routes, but there should also be informal ones which can provide rapid and comprehensive information exhange. If we can have informal advanced international coordination as described by Prof. Pigford, we would be better off in dealing with domestic controversy when such an event occurs. I think universities can play a very important role in that kind of international coordination.

Dr. D. G. H. LATZKO

Let's not forget that in the year 2000, we'll have maybe 20% of our energy coming from nuclear and the remainder will come from coal and oil. Now, that coal will not be used in the way it is being used now. It will take very sophisticated equipment. The oil will come from, or will at least partly come from, very hostile environments. So any safety research, I think, should be directed in such a way as to make those parts of energy development also safe.

If something blows up in the middle of the North Sea, it's as much a disaster as the Three Mile Island, and it may happen. So, again I should like to emphasize that university research should be connected to structural mechanics, for instance, because a piece of piping does not know whether it's in a nuclear station or whether it's at the bottom of the sea, and the one thing we must know is that it doesn't rupture. I think university research is applied to that kind of problem, or, say, to the T-joints in an offshore structure. It would not only have very broad benefits, but can give students broader job opportunities, because Shell Oil, for instance, would be very happy to get people who happened to know a lot about pipe cracks. It can also save university safety research from becoming too academic.

Dr. T. PIGFORD

I think safety research is certainly one of the important fields, and it offers so many beautiful problems. The only part of Prof. Kondo's summary that I would question is that it needs special facilities. So many things we need to know are being studied in simple university laboratories. We don't really know enough about the laws governing the heat flow of two-phase media. We understand it only in some extreme cases. We don't understand the laws governing the rejecting phenomena of a very hot plate, and it's so important. These simple experiments designed to yield understanding are also yielding new design data, much to my surprise. So I think that so many areas are suitable in this.

One danger is that when the problems are brought to us (and to you perhaps) by an agency like NRC (Nuclear Regulatory Commission) or the Electric Power Research Institute, they want a solution in six months. We resist that. We find that if we resist hard enough *they* seek funds in other places, and frankly we get them. We can do it properly. There are so many beautiful problems, which simple laboratories and facilities can treat.

Dr. M. BENEDICT

I want to make a specific comment on the role of safety-related research to nuclear education, and then I would just like to make some general comments about this very fine conference. We have found at MIT, thanks to the pioneering work of my colleague, Prof. N. Rasmussen, that the kind of techniques for probabilistic risk analysis which he introduced in the so-called Rasmussen Report provide a very fertile field for both student education and student participation in on-going research programs, because of the kind of detailed analysis that one has to make of engineering systems and their operating characteristics provides a much more intimate understanding of how some of these complex nuclear facilities work. So safety-related analysis and safety-related research are now an important part of our nuclear engineering curriculum.

Now, if I may just say a few general words about this very fine conference, I particularly want to congratulate the Department of Nuclear Engineering of the University of Tokyo on organizing and presenting such an interesting and informative conference on nuclear engineering topics. To me it's been a sort of miniature Geneva Conference on the Peaceful Uses of Atomic Energy. When Prof. Pigford and I started MIT's program of research and instruction in nuclear engineering back in 1952, we didn't have the faintest idea that one day we'd be attending the twentieth anniversary celebration of our eminent sister university, the University of Tokyo. During these twenty years you've created a great new department at a great university and are obviously playing a prominent role in making Japan one of the leading nations in the peaceful uses of nuclear energy, using it to improve the prosperity of your nation and the quality of life. So I feel certain that your university, universities of the United States, and those elsewhere in the world can continue to make important contributions to both education and research in nuclear engineering and certainly will train the next generation of nuclear engineers who will make nuclear energy the important contributor to the world's energy supply. We all expect that. So I congratulate the Nuclear Engineering Department of the University of Tokyo on the important role you're playing in applying nuclear energy to the future.

Dr. K. OSHIMA

Thank you very much. Since the time is up, I would like to conclude the meeting. Looking at the overall meeting, whose subject is the role of nuclear engineering for an uncertain future, I felt that nuclear energy does have an important role for future energy supply but also that there are a lot of problems to be solved, and because of these problems and uncertainties it is important for us to continue our research. If all the questions are solved, there is no role for us to play. Therefore, uncertainties and problems are a challenge and very important to all of us. As already indicated in the discussion, universities have a very important role to play in training engineers and students, and at the same time universities should be very innovative in leading the industry as well as the society.

There are many aspects that we can devote ourselves to. If one thinks about the problem, one tends to treat it unidimensionally. Therefore, it is important for us to get together from time to time to try to collect our wisdom. And from this point of view, the international symposium as a whole, as well as this panel discussion, can be a starting point for the coming decades.

I believe it is important to continue this sort of international contact or dialogue. The views in our country may be different from those of the United States or Europe. Of course, there is a basic common ground we stand on, but there can be some differences on specific details. So, if we really want to pursue our future with international links, it is important to try to emphasize the commonality of our concerns.

Professor Pigford has been very incisive to say that professors in nuclear engineering should try to give lectures in subjects which are different from their own major fields in order to teach students with sufficient understanding about nuclear engineering; at the same time we are expected to do creative research. I think this is not an easy task to do, and we have to find a good synthesis of these requirements.

Finally, I would like to express my gratitude for the financial support by industry for this seminar so that we could have such prominent scholars and researchers from various parts of the world. This will not be an end in itself but is going to be the starting point for our effort in the coming decades.

Closing Remarks

On behalf of the members of our department, I would, first of all, like to extend my appreciation to all of the participants who have been very conscientious and attentive in making contributions to this discussion. My thanks go to the professors and scholars who have come from the United States and from Europe to give us very valuable presentations.

Twenty-year sounds very long but is not really so compared to the long history of other traditional scientific areas. Nevertheless, our idea in holding this symposium commemorating the 20th anniversary was to try to find ways and means of further development of our department. I think we were able to make the symposium really fruitful and I am indeed very grateful to all the participants. I think this symposium identified what we have done and where we are at present, and I hope that this will be a very important step for us to guide our activities in the future.

Thank you very much.

November 6, 1980

Yoshio ANDO
Vice-chairman
The Organizing Committee

Abbreviations

ACRS: Advisory Committee on Reactor Safeguards (U.S.)
AEC: Atomic Energy Commission
AGR: Advanced Gas-Cooled Reactor
ALARA: As Low As Reasonably Achievable
ASDEX: Axi-Symmetric Divertor Experiment (FRG)
ASME: American Society of Mechanical Engineers
ATR: Advanced Thermal Reactor
AVLIS: Atomic Vapor Laser Isotope Separation Process
AVM: Atelier de Vitrification, Marcoule

BMI: Bundesministerium des Innern
BNFL: British Nuclear Fuels Limited
BWR: Boiling Water Reactor

CAD: Computer-Aided Design
CAM: Computer-Aided Manufacture
CDA: Core-Disruptive Accidents
CDFR: Commercial Demonstration Fast Reactor (U.K.)
CEA: Commissariat à l'Energie Atomique
COGEMA: Compagnie Générale des Matières Nucléaires
CRBR: Clinch River Breeder Reactor
CTC: Cluster Test Coil
CTR: Controlled Thermonuclear Reactor
CVN: Charpy V-Notch

DEMO: Demonstration (Fusion) Reactor (Japan)
DITE: Divertor Injection Tokamak Experiment (U.K.)
DIVA: Divertor Assembly (Japan)
DRS: Deutsche Risiko Studie
DYMAC: Dynamic Material Accountancy

ECCS: Emergency Core Cooling System
ECH: Electron Cyclotron Heating
EDF: Electricité de France
ENEL: Ente Nazionale per l'Energia Elettrica
EPRI: Electric Power Research Institute (U.S.)
ETL: Electrotechnical Laboratory (Japan)

EURODIF: Uranium Enrichment (Diffusion Process) Company in Europe (Shareholders are France (Iran), Italy, Belgium and Spain)

FBR: Fast Breeder Reactor
FBTR: Fast Breeder Test Reactor (India)
FED: Fusion Engineering Device (U.S.)
FFTF: Fast Flux Test Facility (U.S.)
FMIT: Fusion Material Irradiation Test Facility (U.S.)
FSTR: Fuel Safety Test Reactor (Japan)

GCR: Gas-Cooled Reactor

HAW: Highly Active Waste
HBTXIA: High Beta Toroidal Experiment-IA (U.K.)
HWR: Heavy Water (Moderated) Reactor

IAEA: International Atomic Energy Agency
ICRF: Ion Cyclotron Resonance Frequency
ICRP: International Commission on Radiological Protection
IEEE: Institute of Electrical and Electronics Engineers, Inc.
IHX: Intermediate Heat Exchanger
INFCE: International Nuclear Fuel Cycle Evaluation
INPO: Institute of Nuclear Power Operation (U.S.)
INTOR: International Tokamak Reactor
IPPN: Institute of Plasma Physics, Nagoya University
ISI: In-Service Inspection
ISXB: Impurity Study Experiment-B (U.S.)

JAERI: Japan Atomic Energy Research Institute
JET: Joint European Torus
JFT-2: JAERI Fusion Torus-2
JIPP-T II: Japan Institute of Plasma Physics Torus II

KFK: Kernforschung Karlsruhe GmbH
KNK-II: Kompakte Natriumgekühlte Kernreakoranlage-II

LAW: Low-Active (Liquid) Waste
LHH: Lower Hybrid Heating
LIB: Light Ion Beam
LLL: Lawrence Livermore Laboratory
LMFBR: Liquid-Metal-Cooled Fast Breeder Reactor
LOCA: Loss of Coolant Accidents
LWR: Light Water (Moderated) Reactor

MAW: Medium-level Aqueous Waste
MBPD: Million Barrel per Day
MFTF: Mirror Fusion Test Facility (U.S.)

MIT: Massachusetts Institute of Technology
MITI: Ministry of International Trade and Industry (Japan)
MLIS: Molecular Laser Isotope Separation Process
MOX: Uranium-Plutonium Mixed Oxide
MWD/T: Megawatt Day per Metric Ton, Heavy Metal

NBT: Nagoya Bumpy Torus
NDE: Non-Destructive Examination
NERSA: Centrale Nucléaire Européenne à Neutorons Rapides S.A.
NPT: Non-Proliferation Treaty
NRC: Nuclear Regulatory Commission
NSAC: Nuclear Safety Analysis Center (U.S.)
NSRR: Nuclear Safety Research Reactor (Japan)
NSSS: Nuclear Steam Supply System

OECD: Organisation for Economic Co-operation and Development
ORNL: Oak Ridge National Laboratory
OTEC: Ocean Thermal Energy Conversion

PCI: Pellet-Clad Interaction
PCIOMR: Pre-Conditioning Interim Operating Management Recommendation
PCM: Power-Cooling Mismatch
PCT: Peak Clad Temperature
PDX: Poloidal Divertor Experiment (U.S.)
PIE: Post-Irradiation Examination
PLT: Princeton Large Torus
PNC: Power Reactor and Nuclear Fuel Development Corporation (Japan)
PURR: Prototype d'Usine de Retraitement des Rapides
PWR: Pressurized Water Reactor

R&D: Research and Development
RDD & D: Research, Development, Demonstration and Deployment
RFP: Reversed Field Pinch (U.K.)
RIA: Reactivity-Initiated Accident
RSK: Reaktor-Sicherheitskommission
RWE: Rheinisch-Westfälisches Elektrizitätwerk

SMiRT: Structural Mechanics in Reactor Technology
SWU: Separative Work Unit

TASTEX: Tokai Advanced Safeguards Technology Exercise
TEPCO: Tokyo Electric Power Co.
TEXTOR: Torus Experiment for Technology-Oriented Research, OECD/IEA
        (FRG)
TFTR: Tokamak Fusion Test Reactor (U.S.)
TMI: Three Mile Island
TMX: Tandem Mirror Experiment (U.S.)

TNT: Tokyo Noncircular Tokamak
TOR: Traitment d'Oxydes Rapides
TORIUT: Tori, University of Tokyo
TRIAM: Tokamak of Research Institute of Applied Mechanics (Japan)

UCOR: Uranium Enrichment Corporation of South Africa
UKAEA: United Kingdom Atomic Energy Authority
UN: United Nations
URENCO: Uranium Enrichment (Gas-Centrifuge Process) Company in Europe
       (Shareholders are UK, FRG and Netherlands)
USDOE: United States Department of Energy
USNRC: United States Nuclear Regulatory Commission

VAK: Versuchskraftatomwerk Kahl GmbH
VHTR: Very High Temperature Reactor

WAK: Wiederanfarbeitungsanlage
WASTEF: Waste Safety Test Facility (Japan)
WEC: World Energy Conference
WOCA: World Outside Centrally Planned Economies Area

Conference Agenda

(* invited lecture)

November 5

Conference Registration (9:00–9:30 a.m.)

Opening (9:30–9:45 a. m.)

    Opening Address   Prof. Keichi OSHIMA
                                  Chairman,
                                  The Organizing Committee

    Opening Remarks   Prof. Yoshihiro HISAMATSU
                                  Dean, Faculty of Engineering
                                  University of Tokyo

Part I: "International Cooperation on Energy" (9:45–11:15 a.m.)

    Co-chairmen: Prof. Keichi OSHIMA
                                    Department of Nuclear Engineering
                                    University of Tokyo

                                    Prof. Mamoru AKIYAMA
                                    Department of Nuclear Engineering
                                    University of Tokyo

    *Lecture 1:   "International Cooperation on Nuclear Energy"
                                    Dr. W. Kenneth DAVIS
                                    Vice President,
                                    Bechtel Power Corp.

    Lecture 2:   "Nuclear Energy and International Cooperation"
                                    Prof. Keichi OSHIMA

Luncheon (11:15 a.m. – 1:00 p.m.)

Part II: "Toward an Acceptable Fuel Cycle Scheme" (1:00 – 3:00 p.m.)

    Co-chairmen: Prof. Takaaki TAMURA
                                    Nuclear Engineering Research Laboratory
                                    University of Tokyo

                                    Prof. Ryohei KIYOSE
                                    Department of Nuclear Engineering
                                    University of Tokyo

    *Lecture 1:   "The Nuclear Fuel Cycle: An Overview"

Prof. Manson BENEDICT
Institute Professor Emeritus,
Massachusetts Institute of Technology

*Lecture 2: "Reprocessing Technology in Europe"
Dr. Cyril BUCK
Former Director,
Reprocessing Division
British Nuclear Fuels Limited

Lecture 3: "Management of the Nuclear Fuel Cycle"
Prof. Ryohei KIYOSE

Coffee Break (3:00 – 3:15 p.m.)

Part III: "Engineering Philosophy on Safety" (3:15 – 5:15 p.m.)

Co-chairmen: Prof. Yoshitsugu MISHIMA
Department of Nuclear Engineering
University of Tokyo

Prof. Shiori ISHINO
Department of Nuclear Engineering
University of Tokyo

*Lecture 1: "The Management of Nuclear Safety: Lessons Learned
from the Accident at Three Mile Island"
Prof. Thomas PIGFORD
Department of Nuclear Engineering
University of California, Berkeley

*Lecture 2: "Nuclear Safety: Its Achievement in Perspective"
Dipl.-Phys. Armin JAHNS
Geschäfsführer
Reaktorsicherheitskommission (RSK)

Lecture 3: "Safety-Oriented Research on Power Reactor Fuels"
Prof. Yoshitsugu MISHIMA

November 6

Part IV: "Transition to Fast Breeder Reactors" (9:00 – 10:30 a.m.)

Co-chairmen: Prof. Akira OYAMA
Director,
Power Reactor and Nuclear Fuel Corp.

Prof. Akira SEKIGUCHI
Department of Nuclear Engineering
University of Tokyo

*Lecture 1: "A Breeder Strategy to Solve Uncertain Future Problems in Energy Supply: The French Example"
Dr. Jean F. PETIT
Coordinator, Fast Reactor Projects
Commissariat a l'Energie Atomique

Lecture 2: "RDD & D on the Fast Breeder Reactor"
Prof. Shigehiro AN
Nuclear Engineering Research Laboratory
University of Tokyo

Part V: "Research and Development of Fusion Technologies" (10:30 a.m. – noon)

Co-chairmen: Prof. Osamu NISHINO
Professor Emeritus,
University of Tokyo

Prof. Masayoshi KANNO
Department of Nuclear Engineering
University of Tokyo

*Lecture 1: "Nuclear Fusion Research: Status and Prospects"
Dr. Hugh A. B. BODIN
Culham Laboratory
United Kingdom Atomic Energy Authority

Lecture 2: "Japan Fusion Research Activities and Future Plans"
Prof. Taijiro UCHIDA
Department of Nuclear Engineering
University of Tokyo

Luncheon (noon–1:30 p.m.)

Part VI: "Nuclear Engineering and Technological Innovation" (1:30–3:00 p.m.)

Co-chairmen: Prof. Yoshio ANDO
Department of Nuclear Engineering
University of Tokyo

Prof. Yoichi TAKAHASHI
Nuclear Engineering Research Laboratory
University of Tokyo

*Lecture 1: "The Impact of Nuclear Engineering on Technological Improvements in Other Fields"
Prof. D. G. H. LATZKO
Department of Mechanical Engineering
Delft University of Technology

Lecture 2: "Soft Energy vs Nuclear Energy"
Prof. Yoshio ANDO

Coffee Break (3:00 – 3:15 p.m.)

Part VII: "Panel Discussion: The Role of Nuclear Engineering Research and Education for Energy Futures" (3:15–5:15 p.m.)

Moderator: Prof. Keichi OSHIMA

Panelists: Prof. Manson BENEDICT

Prof. Shunsuke KONDO
Department of Nuclear Engineering
University of Tokyo

Prof. D. G. H. LATZKO

Prof. Thomas PIGFORD

Prof. Akira SEKIGUCHI

Prof. Yoichi TAKAHASHI

Closing (5:15 – 5:20 p.m.)

Clossing Address   Prof. Yoshio ANDO
Vice-chairman,
The Organizing Committee

Index of Lecturers

Shigehiro AN

Professor An graduated from the Department of Physics, Faculty of Science, University of Tokyo, in 1948. He was appointed lecturer in the Department of Nuclear Engineering, Faculty of Engineering, University of Tokyo, in 1961, and then an associate professor in the same department in 1963. In 1967, he became Professor in the Nuclear Engineering Research Laboratory, Faculty of Engineering, University of Tokyo. He served as Chairman of the Research Laboratory from 1967 to 1974 and from 1977 to 1980.

His special field is reactor engineering, especially fast breeder core design and safety. Energy systems analysis and fusion reactor design are also his interests in recent years. He was in CEN de Cadarache, France, in 1964–1965, working on the French Fast Breeder Reactor Project. He was in charge of construction of the Fast Neutron Source Reactor at the University of Tokyo, which went on-stream in 1971.

He has been active in nuclear energy research and development in connection with the Ministry of Education, Science and Culture, Ministry of Foreign Affairs, Science and Technology Agency, Japan Atomic Energy Commission, Japan Atomic Energy Research Institute, Japan Atomic Energy Society, Japan Atomic Industrial Forum, Japan Scientific Council, Power Reactor and Nuclear Fuel Development Corporation. He has been deeply involved in the fast breeder reactor development program in Japan. He is a technical advisor and a chairman of the specialists' committee on fast reactor core design and safety analysis of the PNC.

Yoshio ANDO

Professor Ando was born in 1922 in Tsingtao, China. He graduated from the Department of Naval Architecture, University of Tokyo, in 1945 and was appointed as Lecturer in the same department in 1945. He was promoted to Associate Professor of Naval Architecture in 1948 and appointed Associate Professor of Welded Structure at the Institute of Industrial Science, University of Tokyo, in 1951.

He received his doctorial degree in engineering from the University of Tokyo in 1960.

When the Department of Nuclear Engineering was founded in the University of Tokyo in 1962, he joined it as Professor of Nuclear Structural Engineering and Nuclear Propulsion.

He is a Director or former Director of the Japan Atomic Energy Society, Japan Welding Society, Japan Welding Engineering Society, Japan High Pressure Institute, Japan Nuclear Safety Research Association, Japan Atomic Industrial Forum, Japan Power Plant Inspection Institute, and Thermal and Nuclear Power Association.

He is a member or a chairman of various committee organized by the government such as the Committee on Examination of Reactor Safety, N.S. Mutsu Repair and Review Committee, Committee on Nuclear Structural Safety, Committee on the Study of Stress Corrosion Cracking, and so on.

He is the author or co-author of over 300 publications which focus primarily on nuclear structural engineering both in Japan and worldwide.

He is a Foreign Associate of the National Academy of Engineering of the U.S.A.

Manson BENEDICT

Dr. Manson Benedict is Institute Professor Emeritus at Massachusetts Institute of Technology, where he served as the first head of the Nuclear Engineering Department.

In 1951, Dr. Benedict came to M.I.T. to organize a program of research and instruction in nuclear engineering, and the Department of Nuclear Engineering was established under his leadership in 1958. At M.I.T., Dr. Benedict's principal fields of research have been in isotope separation and fuel cycles for nuclear reactors. He has done extensive research in separation processes, including extraction and azeotropic distillation, the theory of multi-stage operations, and processes for uranium enrichment and heavy water production.

A former member (1958–1968) and past chairman (1962–64) of the General Advisory Committee of the Atomic Energy Commission, and a former member of the Advisory Committee on Reactor Safeguards, he has played a prominent advisory role in the Government's nuclear energy program. He has been scientific advisor to the U.S. delegation to three International Conferences on the Peaceful Uses of Atomic Energy, sponsored by the United Nations in Geneva in 1955, 1958 and 1964.

Between 1935 and 1951, Dr. Benedict conducted extensive research in physical chemistry while a student at M.I.T., and later at Harvard, the M. W. Kellogg Company, and Hydrocarbon Research, Inc. His work at M.I.T. concerned the absolute temperature scale; at Harvard he concentrated on the properties of gases at high pressures and the properties of aqueous solutions at high pressures. At the M. W. Kellogg Company, he developed a widely used equation of state for hydrocarbons.

During World War II, Dr. Benedict was the head of the process development division of the Kellex Corporation and was in charge of the process design of the gaseous diffusion plant for the concentration of uranium-235 for the Manhattan Project at Oak Ridge.

In 1966, Dr. Benedict received the Perkin Medal of the American Section of the Society of Chemical Industry. The Society cited his contributions to the successful design and operation of the first gaseous diffusion process. He was given the 1963 American Chemical Society Award in Industrial and Engineering Chemistry for his successful work in physical chemistry, petroleum process engineering, isotope separation processes and nuclear engineering. From the American Institute of Chemical Engineers he has received the William H. Walker Award and the Founders Award. The Atomic Energy Commission has given Dr. Benedict the AEC Citation and, in 1972, the Enrico Fermi Award. He received the National Medal of Science in 1975.

Born in Lake Linden, Michigan, in 1907, Dr. Benedict received the B. Chem. from Cornell in 1928 and the Ph. D. in Chemistry from M.I.T. in 1935.

Dr. Benedict is a member of the National Academy of Science and the National Academy of Engineering. He was president of the American Nuclear Society from 1962–63, and he is a past director of the American Institute of Chemical Engineers and the Atomic Industrial Forum.

Dr. Benedict and his wife, the former Marjorie Oliver Allen, met while both were working for their Ph. D's in physical chemistry at M.I.T. Married in 1935, they have two daughters. They live at 25 Byron Road in Weston, Massachusetts.

Hugh A. B. BODIN

Hugh A. B. Bodin studied at the University of Glasgow where he obtained a BSc in natural philosophy followed by a Ph. D. He then joined the United Kingdom Atomic Energy Authority in 1955 at AWRE Aldermaston, helping to set up a new group to study controlled fusion research, on which he has worked ever since. His early work was on pulsed high beta plasmas including fast linear pinches and theta pinches. He joined Culham Laboratory when it was set up in 1962 and since then much of his work has been on high beta toroidal pinches and he now heads work on the reversed field pinch. His publications include many papers on diffusion and instabilities in high beta plasma, general reviews, and contributions to textbooks. He has served on a number of national and international committees on fusion research and acted in an advisory capacity in developing countries. He is an adjunct professor at Texas Technical University.

Cyril BUCK

After training in the chemical industry Cyril Buck entered the atomic energy field in 1946. His earlier responsibilities included design and construction of major plants at Springfields for the purification of uranium and the manufacture of reactor fuel elements, and at Windscale for design and construction of the first primary separation plant for the extraction of plutonium and associated plants for dealing with nuclear wastes. In 1962 he was appointed Director, Chemical Plant Design in the Engineering Group of UKAEA, and was responsible for all engineering activities including development, design and construction of processing plants at the three factories of the UKAEA Production Group, namely Capenhurst, Springfields and Windscale. This work included the second separation plant at Windscale and also the chemical separation plant at Dounreay. On the formation of British Nuclear Fuels Limited in 1971 he was appointed Director of Plant Design and appointed a member of the Main Board of the company in 1973.

In 1974 the company was re-organized on the basis of three operational divisions and he was appointed Director of the Reprocessing Division, having responsibility for the total activities associated with the reprocessing and plutonium fuel manufacturing facilities of the company, namely, policy, finance, commercial matters, research and development and engineering. He retired from the company in 1978, and since that time as a consultant has represented U.K. interests in the International Fuel Cycle Evaluation Study. In particular acting as Co-Chairman with Dr. S. Tamiya and Dr. W. Marshall on the Working Group dealing with reprocessing and plutonium fuel management, he served as the UKAEA delegate on other working groups dealing with enrichment and fuel management. He is now Chairman of the Technical Sub-Group of the IAEA Expert Group on International Spent Fuel Management.

W. Kenneth Davis

Mr. W. Kenneth Davis is a vice president of Bechtel Power Corporation, responsible for planning and advanced development. He has extensive experience and expertise in nuclear power technology and its application as well as energy developments generally. Mr. Davis is currently vice president of the National Academy of Engineering and vice president and president-elect of the American Institute of Chemical Engineers. He is also Adjunct Professor of Engineering and Applied Science at the University of California, Los Angeles. He is active in many of the energy/engineering societies of the United States and has recently served as chairman of the Supply/Delivery Panel for the Committee on Nuclear and Alternative Energy Sources for the U.S. National Academies of Sciences and Engineering and as a U.S. participant in the World Coal Study. He is a director of the U.S. National Committee of the World Energy Conference and is second vice chairman for the current year.

Mr. Davis has BS and MS degrees in chemical engineering from MIT. He is the recipient of the Arthur S. Flemming Award, 1956; the AIChE Professional Progress Award, 1958; and the AIChE Robert E. Wilson Award, 1969.

Before joining Bechtel, Mr. Davis was Director of Reactor Development of the Atomic Energy Commission, Manager of Research at California Research and Development Company, Professor of Engineering at UCLA, and had previous experience in oil refinery process design.

Mr. Davis is the author or co-author of over 200 publications which focus primarily on nuclear energy (its history, its present status, and its future) and energy in general, both for the United States and worldwide.

Armin JAHNS

Executive Secretary, Reactor Safety Commission (RSK), Director of RSK-Office, Gesellschaft für Reaktorsicherheit (GRS) mbH in Köln, Federal Republic of Germany.

University of Göttingen 1958-1962. Free University of Berlin (West) 1962-1964. Graduated as Diplom-Physicist. Hahn-Meitner-Institute for Nuclear Research 1964-1965, worked on low-energy physics and radiation physics. Joined the Institute for Reactor Safety (IRS) in Köln in 1965. 1966-1967 National Reactor Testing Station (now INEL) in Idaho, U.S.A.; Special Power Excursion Reactor Test (SPERT)-Program on an exchange agreement between AEC and FRG. 1967-1971 reactor safety analyses for light water reactors and fast breeder reactors in licensing procedures and project management for the SNR-300 safety assessment. Present position since 1971. Stayed with GRS when the company was formed in 1977 by IRS-Köln and LRA-München.

Memberships:
Kerntechnische Gesellschaft (KTG), American Nuclear Society (ANS).

Ryohei KIYOSE

Dr. Ryohei Kiyose is Professor of Nuclear Chemical Engineering at the Department of Nuclear Engineering, in the Faculty of Engineering of the University of Tokyo. Dr. Kiyose is now serving as the chairman of the Graduate Study Course of Nuclear Engineering for the Graduate School of Engineering at the university. He received a BS in Physics from the University of Tokyo in 1952, finished his graduate course in Chemical Engineering in 1954, and after working as a research associate at the Department of Chemical Engineering, joined the Department of Nuclear Engineering as a lecturer in 1958. He has served as an associate professor at the department from 1961 to 1976, and as a professor since 1976. The title of his doctoral thesis written in 1975 was "Systems Analysis and Optimization Studies on Nuclear Fuel Cycle Processes". Dr. Kiyose is now a member of the Board of Directors of the Atomic Energy Society of Japan, and also is serving as a member of several advisory committees on safety review, safety standards, and research and development plans of nuclear technologies for the Japanese government. He received the Exceptional Service Award in 1980 from the American Nuclear Society on the occasion of its 25th anniversary. His fields of interest have been in-core fuel management, safety assessment of nuclear fuel facilities including criticality safety, radioactive waste management, and other related subjects.

D. G. H. Latzko

Degree: M. Sc. in Mechanical Engineering, Delft University of Technology, 1949.

Positions held: Mechanical development engineer, AKU N.V. (now: AKZO), Arnhem, 1949–1952. Design engineer, subsequently Assistant Manager, steam turbine department, Werkspoor N. V., Amsterdam, 1952–1957. Chief Mechanical Engineer, P.E.N. (electric utility), Bloemendaal, 1957–1965. Adjunct Professor of Mechanical Engineering, Delft University of Technology, 1961–1965. Professor of Mechanical Engineering, subsequently head, Laboratory for Thermal Power Engineering, Delft University of Technology, 1965-present.

Memberships, offices held (limited list): Member, Royal Netherlands Academy of Science. Member, past chairman, Euratom Scientific and Technical committee. Member, Netherlands General Energy Council (AER). Vice chairman, Netherlands Energy Research Council. Member, Netherlands Reactor Safety Committee. Chairman, International Working Group on Reliability of Reactor Pressure Components (IWG-RRPC), IAEA, Vienna.

Publications: Author of over 30 papers in scientific journals and conference proceedings. Invited speaker for the first Robert D. Wylie Memorial Lecture, 4th International Conference on Pressure Vessel Technology, London, 1980. Co-author and editor of: *"Post-Yield Fracture Mechanics,"* Applied Science Publishers, Barking, England (1979).

Yoshitsugu Mishima

Dr. Mishima has been Professor of Nuclear Metallurgy in the Department of Nuclear Engineering, Faculty of Engineering, University of Tokyo, since 1963.

He graduated from the University in 1944 in the course of Metallurgy and was appointed Associate Professor in the Engineering Research Institute, Faculty of Engineering, in 1949 after his post-graduate course of 5 years.

His doctoral thesis was on the ageing phenomena in metallic materials; the first part was presented to the First World Metallurgical Congress in Cleveland (U.S.A.) in 1951.

Dr. Mishima entered the nuclear metallurgy field in the early 1950s, and his special field has been the study of zirconium, beryllium and other fuel cladding materials.

Since 1960, he has been a leader in safety-related fuel study in Japan; for example, he is the chairman of the Expert Committee on Fuel Safety (NEN-ANSEN).

Dr. Mishima has been a member of the Japanese Committee on the Examination of Reactor Safety for the past 20 years and has been its chairman since 1979. He is the Technological Advisor to JAERI on VHTR Project and Fuel Study and was the chairman of the LMFBR Planning Committee of the Japanese AEC in early 1960s. He is now the chairman of the Expert Committee on Fuel and Materials for ATR as well as those for FBR in PNC. He also has been the chairman of the Advisory Fuel Cycle Committee for MITI since 1968.

Dr. Mishima has been the official representative of Japan to the Fuel Specialists Committees of both IAEA (IWG) and OECD (NEA).

Keichi OSHIMA

Professor Oshima has been Professor of Radiation and Reactor Chemistry at the Department of Nuclear Engineering, the University of Tokyo since 1961. Born in Tokyo in 1921, Dr. Oshima graduated in Applied Chemistry from the University of Tokyo in 1944 and received his degree of Doctor of Engineering from the same university in 1959. Before being appointed to his present position, he was an Associate Professor of Physical Chemistry at the Institute of Science and Technology from 1950 and Associate Professor of Cryogenic Engineering at the Institute for Solid State Physics of the University of Tokyo from 1958.

Besides his accademic studies in nuclear engineering, he has been engaged in studies on technology policy as well as energy problems. He has served as the director for Science, Technology and Industry, OECD, on leave from the University (1974–1976). In 1980, Dr. Oshima was appointed as a foreign member of the Royal Swedish Academy of Engineering Science (IVA).

He works for the Japanese Government as a member of the advisory committees to the Ministry of International Trade and Industry, the Science and Technology Agency, the Prime Minister's Office and other agencies. He has been a member of the UNIDO Consultative Group on Appropriate Industrial Technology since 1977.

Jean F. PETIT

Jean F. Petit was born in April 1928. He graduated from French Air Military Academy in 1951 and served as an officer fighter pilot in the French Air Force from 1951 to 1966, while graduating in Nuclear Physics and Engineering in 1963 from the National Institute of Nuclear Sciences and Engineering at Saclay.

He joined CEA in 1966 in the Reactor Safety Division, where he was in 1970 in charge of advanced reactors safety analysis (breeders and water Reactors for submarines). He was assigned as the head of this division in 1973. In 1978, he was nominated Fast Reactor Projects Coordinator of CEA. He is in this function, responsible for R & D budgets and programs, for relations with national industry and utility (EDF), and of international cooperation.

Thomas H. PIGFORD

Thomas H. Pigford is Professor of Nuclear Engineering at the University of California at Berkeley, California. He specializes in the fuel cycle for nuclear power reactors, radioactive waste management, nuclear reactor design, and nuclear safety. He joined the University of California in 1959 as the first Chairman of the Department of Nuclear Engineering.

Professor Pigford received his doctoral degree in Chemical Engineering from the Massachusetts Institute of Technology. He was Assistant and Associate Professor of Chemical and Nuclear Engineering at M.I.T. He was director of the M.I.T. Graduate School of Engineering Practice at Oak Ridge, Tennessee. With Prof. Manson Benedict, he organized the graduate program in Nuclear Engineering at M.I.T.

In 1957 Prof. Pigford participated in the organization of the General Atomic Company, in San Diego, California, where he was Director of Engineering and Director of Nuclear Reactor Projects. The Marine Gas-Cooled Reactor and the High-Temperature Gas-Cooled Reactor were projects initiated under his direction.

Professor Pigford is a member, former Director, and Fellow of the American Nuclear Society. He received the Society's Arthur H. Compton Award for his work in nuclear engineering education and reactor safety. He is a member of the American Institute of Chemical Engineers, and he has recently been named recipient of the Robert E. Wilson Award. He is an elected member of the National Academy of Engineering. He is a member of the National Academy's Committee on Radioactive Waste Management and is Chairman of its Panel on the Geologic Isolation of High-Level Radioactive Waste.

In 1975 Prof. Pigford was Visiting Professor at Kyoto University in Japan and recipient of a fellowship from the Japan Society for the Promotion of Science.

Professor Pigford was a member of the President's Commission to Investigate the Accident at Three Mile Island. He is now Chairman of the Advisory Council for the Institute of Nuclear Power Operations for the United States electric utility industry. He is a consultant to government and industry.

Taijiro UCHIDA

Professor Uchida is one of the pioneers of fusion research in Japan. He graduated from the University of Tokyo in 1953 and entered the plasma-fusion research field in 1957, after working 4 years in private companies. His career of 24 years includes 6 years at Nihon University, 10 years at Nagoya University and 8 years at the University of Tokyo. He received a D. Sc. in Engineering in 1962 and now a Professor at the University of Tokyo. He also serves as a science adviser to the Ministry of Education, Science and Culture.

In Japan, he is a member of the Nuclear Fusion Council under the Japan Atomic Energy Commission, which functions as the coordinating body for all national fusion research projects, and a member of the Science Council's Fusion Committee which promotes fusion research activities in universities. Recently he has contributed to two future fusion R & D plans: one for national laboratories and the other for universities.

Internationally, he is a member of the Fusion Power Coordination Committee of the International Energy Agency (IEA) and a member of the IEA's Executive

Committee on TEXTOR cooperation. He is also playing an active//part in bilateral cooperation with the U.S.A. and the U.S.S.R.

Professor Uchida is the chairman of a national steering committee on grants-in-aid for fusion research and a member of the steering committee of the Institute of Plasma Physics at Nagoya University. He is also a member of the selection committee of the Japan Society for the Promotion of Science.
Memberships:

Japan Nuclear Society, Japan Physical Society, American Nuclear Society, American Physical Society, Institute of Electrical Engineers of Japan, Laser Society of Japan, High Temperature Society.

Publications: more than 30 papers in scientific journals and conference proceedings; general publications, textbooks, and translations.